It's All a Matter of Time: Exploring the Past, Present and Future of Artworks through Dating, Authentication, Material Characterization, Monitoring and Restoration

It's All a Matter of Time: Exploring the Past, Present and Future of Artworks through Dating, Authentication, Material Characterization, Monitoring and Restoration

Guest Editors

Letizia Bonizzoni
Valeria Comite
Anna Galli
Rosina Celeste Ponterio

MDPI

Basel • Beijing • Wuhan • Barcelona • Belgrade • Novi Sad • Cluj • Manchester

Guest Editors
Letizia Bonizzoni
Dipartimento di Fisica Aldo Pontremoli
University of Milan
Milan
Italy

Valeria Comite
Dipartimento di Chimica
University of Milan
Milan
Italy

Anna Galli
Dipartimento di Scienza dei Materiali
University of Milano-Bicocca
Milan
Italy

Rosina Celeste Ponterio
Consiglio Nazionale delle Ricerche—Istituto per i Processi Chimico-Fisici
Messina
Italy

Editorial Office
MDPI AG
Grosspeteranlage 5
4052 Basel, Switzerland

This is a reprint of the Special Issue, published open access by the journal *Heritage* (ISSN 2571-9408), freely accessible at: https://www.mdpi.com/journal/heritage/special_issues/2N5HI35C96.

For citation purposes, cite each article independently as indicated on the article page online and as indicated below:

Lastname, A.A.; Lastname, B.B. Article Title. *Journal Name* **Year**, *Volume Number*, Page Range.

ISBN 978-3-7258-7092-9 (Hbk)
ISBN 978-3-7258-7093-6 (PDF)
https://doi.org/10.3390/books978-3-7258-7093-6

Contents

heritage

MDPI

Article

Non-Invasive and Micro-Invasive Analyses for Supporting the Attribution of Three 17th Century Wooden Crucifixes: Pictorial Materials and Construction Techniques of *Fra' Umile da Petralia*

Maria Francesca Alberghina [1], Salvatore Schiavone [1,*], Antonio Alaimo [2], Enza Gulino [2], Giuseppe Mantella [3], Luciana Randazzo [4,5], Michela Ricca [6], Valeria Comite [7,*], Silvestro Antonio Ruffolo [6] and Mauro Francesco La Russa [6]

1 S.T.Art-Test di S. Schiavone & C Sas, Via Stovigliai 88, 93015 Niscemi, Italy; francesca.alberghina@gmail.com
2 Alaimo e Gulino Conservazione e Restauro, via Trinità, 2/A, 90024 Gangi, Italy; alaimogulino@libero.it (A.A.); alaimogulino@gmail.com (E.G.)
3 Giuseppe Mantella Restauro Opere D'ARTE, Via Circonvallazione Paparo 25, 88060 Isca sullo Ionio, Italy; mantellag@yahoo.it
4 Department of Earth and Marine Sciences (DiSTeM), University of Palermo, Via Archirafi 26, 90123 Palermo, Italy; luciana.randazzo@unipa.it
5 Istituto Nazionale di Geofisica e Vulcanologia (INGV), 90146 Palermo, Italy
6 Department of Biology, Ecology and Earth Sciences (DiBEST), University of Calabria, Ponte P. Bucci, Cubo 12b, 87036 II Piano, Italy; michela.ricca@unical.it (M.R.); silvestro.ruffolo@unical.it (S.A.R.); mauro.larussa@unical.it (M.F.L.R.)
7 Dipartimento di Chimica, Università Degli Studi di Milano, Via Golgi 19, 20133 Milano, Italy
* Correspondence: info@start-test.it (S.S.); valeria.comite@unimi.it (V.C.)

Abstract

This research deals with a multi-analytical approach to characterise three polychrome wooden crucifixes attributed to *Fra' Umile da Petralia* or, in one case, close to his style. The analysed wooden sculptures belong to (1) the Sanctuary of Saint Umile from Bisignano (Cosenza, Italy); (2) the Santissimo Crocifisso Church at Cutro (Crotone, Italy); (3) the Saint Salvatore Church at Gangi (Palermo, Italy). *Fra' Umile* (Giovanni Francesco Pintorno) was born in Petralia Soprana (Palermo, Sicily) between 1600 and 1601, died on 9 February 1639, and belonged to the Order of Friars Minor. Systematic research of the materials and constructive techniques of wooden sculptures is still not very wide, although diagnostic analyses could represent a useful tool for art historians and restorers for these typologies of works of art. The wooden sculptures were subjected to both non-invasive and micro-invasive investigations. Digital direct X-ray radiography, SEM-EDX, and X-ray fluorescence analyses have been carried out, revealing for the three analysed sculptures a similar construction technique and similar pictorial materials. The provided diagnostic evidence supports the coincident chronology and the same attribution to *Fra' Umile da Petralia* or his workshop, confirming the proposals based on the stylistic comparative analyses and archival historical documents. The results of this first multi-analytical investigation, documenting the artistic technique and construction system, represent a starting point for future systematic study on the artistic production of this prolific artist and sculptor never studied from a technical–scientific point of view to date.

Keywords: multi-analytical investigation; digital radiography; wooden sculptures

1. Introduction

Fra' Umile da Petralia, born Giovanni Francesco Pintorno, was a Sicilian sculptor who left an indelible mark on the Baroque period through his religious artwork. Originating from the small town of Petralia Soprana (Palermo, Sicily) around 1600, his life and artistic production were profoundly shaped by his commitment to the Franciscan Order [1–3]. Very little is known about his early life, but it is assumed that he grew up surrounded by the artisanal traditions of his region, which likely increased his interest in sculpture. At a young age, he joined the Franciscan convent and adopted the name "*Umile*", meaning "humble", a reflection of his spiritual devotion. His sculptures, particularly his wooden crucifixes, were deeply influenced by the Franciscan ideals of simplicity, humbleness, and direct religious experience [2]. His artistic production is known for its emotional depth and naturalism, characteristics that align with the broader themes of the Baroque style, which aimed to evoke strong emotional responses, particularly in religious art. The sculptures attributed to *Fra' Umile da Petralia* often depict Christ in moments of extreme suffering, capturing the raw agony of the Passion with meticulous anatomical detail. The naturalism in his representations of Christ on the cross, showing strained muscles, visible ribs, and pained facial expressions, was designed to inspire viewers to contemplate the suffering of Christ on a personal level [1,2]. His artworks can be found in several churches and convents in Sicily and Southern Italy, indicating his widespread influence during his lifetime.

A systematic search of the pictorial materials and construction techniques of these artworks has rarely been conducted in the scientific community. However, there has been growing interest in the last decade, and several studies describe wooden materials and their distribution in different geographical areas [4–8]. A detailed study of the polychromies wooden statues dated in the 16th and 17th century would allow the correct identification and differentiation of the workshops and active artists and consequently a scientific attribution in addition to that based on artistic–stylistic criteria. This traditional approach is often influenced by the difficult reading of the original models and surfaces altered by the frequent pictorial overlays and remakes to which these wooden sculptures have generally been subjected, due to their continuity of use for religious devotion. These limits can be overcome by systematic radiographic investigation. Diagnostic analysis, referring to both non-invasive and micro-destructive techniques, could unveil information about the materials and the execution techniques [9–15]. In this research, three wooden sculptures attributed to *Fra' Umile da Petralia* were the object of a multi-analytical investigation, coming respectively from (1) the Sanctuary of Saint Umile from Bisignano (Cosenza, Calabria), (2) the Santissimo Crocifisso Church at Cutro (Crotone), and (3) the Saint Salvatore Church at Gangi (Palermo).

The three sculptures investigated were analysed according to a multi-analytical approach based on integrated non-invasive techniques (digital X-ray radiography, UV fluorescence imaging, and portable X-ray fluorescence spectrometry) to understand the state of conservation, to document the construction system and the number of pieces assembled to obtain the whole volume, and to identify the original painting materials. In two cases (Bisignano and Cutro crucifixes), it was possible to deepen the study of the pictorial materials, through scanning electron microscopy analysis and microprobe analysis (SEM-EDS) for the compositional characterisation and stratigraphic study of the pictorial and preparation layers.

Iconographic Study and Stylistic Analysis of the Investigated Wooden Crucifixes

The Crucified Christ from the Sanctuary of *Sant'Umile da Bisignano* at Bisignano (Cosenza, Calabria) (Figure 1a) is attributed to *Fra' Umile da Petralia* and dates back to the

17th century. On the back of the wooden cross, the following inscription is engraved: "1637. P.F. GREGORIO A BISIN. CUSTOD F. HUMILIS A PETRALIA REFOR. SCULP".

Figure 1. Geographic location of the three analysed polychrome wooden sculptures: (**a**) Crucified Christ, Sanctuary of *Sant'Umile da Bisignano* at Bisignano (Cosenza, Calabria), with a detail of the torso and face; (**b**) Crucified Christ, *Santissimo Crocifisso* Church at Cutro (Crotone, Calabria), with a detail of the face and torso; (**c**) Crucified Christ, *Saint Salvatore* Church at Gangi (Palermo, Sicily), with a detail of the torso and face.

The Crucified Christ from *Santissimo Crocifisso* Church at Cutro (Crotone, Calabria) (Figure 1b) is also traditionally attributed to the sphere of *Fra' Umile* [3]. The Jesuit biographer P.G. Macaluso [1,2] places the creation of the Crucified Christ of Cutro between 1636 and 1637, simultaneously with the Crucifixes from Bisignano and the Crucifixes from Polla (Salerno, Campania), with which it shares several details. However, it is distinguished from these and other wooden crucifixes attributed to the same workshop by the presence of a fine artistic detail, a tear falling from the left eye and suspended on the tip of the nose (see detail of the face, Figure 1b).

The Crucified Christ in the Saint Salvatore Church at Gangi (Palermo, Sicily) (Figure 1c) dates back to between the 16th and 17th centuries. This artwork bears no inscriptions, but its stylistic characteristics lead to its attribution to *Fra' Umile da Petralia*. Oral tradition over the centuries reports this as a unique attribution; although no document has ever confirmed this hypothesis, several stylistic elements refer to the style of Giovanni Pintorno or his workshop [3]. Furthermore, the discovery of a document dated in 1656 testifies to the remaking of the pictorial layers by the monks of the nearby town Petralia Sottana (Palermo), custodians and successors of the Franciscan art typical of *Fra Umile da Petralia*. This archival source belongs to the "Accounting book of the Church of the Saint Salvatore from 1646 to 1691" preserved at the Mother Church of Gangi. In fact, in 1656 a payment of "14 *onze*" to the monks is recorded for repainting the wooden sculpture of the SS. Crucifix.

This document therefore demonstrates that the Crucifix was created before 1656, probably according to a style based on naturalism and strong emotional impact spread by *Fra' Umile*, and on that occasion, it was stylistically adapted to the artistic and cultural taste of that time. It is clear that almost twenty years after his death (1639), his realistic and emotional style carried on by the workshop was widespread and preferred by religious commissions.

2. Materials and Methods

The pictorial compositional and stratigraphic characterisation and construction technique study of the polychrome wooden sculptures was carried out with a multi-analytical approach, using both non-invasive and micro-invasive methods. The latter were applied only to two of the crucifixes, the one from the Church of the Santissimo Crocifisso Church at Cutro and the one from the Sanctuary of Saint Umile from Bisignano. The micro-sampling was not authorised for the third crucifix, from the Saint Salvatore Church in Gangi (Palermo). For the latter, only non-invasive elemental analyses were carried out by p-XRF investigation. The study aimed to characterise the artistic technique and the artist's modus operandi in the production of these polychrome wooden sculptures, to attribute or validate the attribution to *Fra' Umile da Petralia*, and to use the information obtained for future comparisons with other artworks attributed to the same artist or his workshop. In addition, studies were carried out to characterise the original layers of the pictorial layers, which provide essential information for restorers during the conservation phases. Finally, considering the minimum number of samples that could be taken, preliminary information was obtained for the identification of the wooden species used for the related part of the statues.

2.1. Micro-Fragment Sampling from the Pictorial Surface

From the Crucifix of the Santissimo Crocifisso Church in Cutro, three micro-fragments were taken from the pictorial layer on both the front and back of the Crucifix, along with an additional sample for wooden species identification.

Moreover, three micro-fragments were taken from the Crucifix of the Sanctuary of *Sant'Umile da Bisignano*: two samples were for the pictorial layer analyses, and one sample was for the preliminary identification of the wood species. Figures 2 and 3 localise and describe the samples from these two analysed polychrome sculptures.

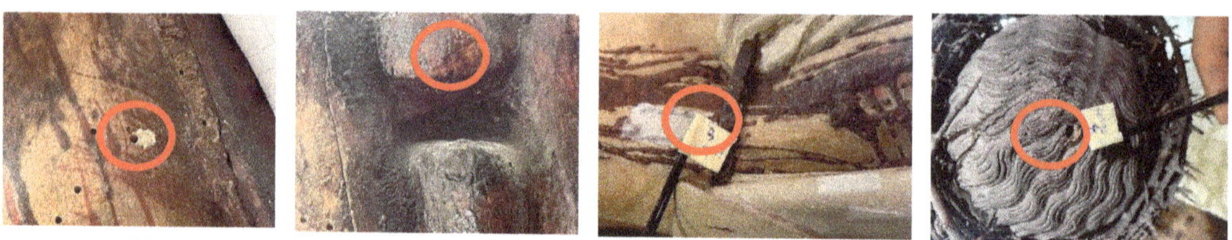

(**a**) CC_01 (**b**) CC_02 (**c**) CC_03 (**d**) CC_04

Figure 2. List of micro-fragments sampled from the Crucified Christ at Cutro: (**a**) CC_01, from flesh tone pictorial layer from the back of the Christ (left side); (**b**) CC_02 from flesh tone pictorial layer from the spinal column; (**c**) CC_03 from dark red pictorial layer from dripping blood; (**d**) CC_04, wood fragment from a lacuna on the top of head.

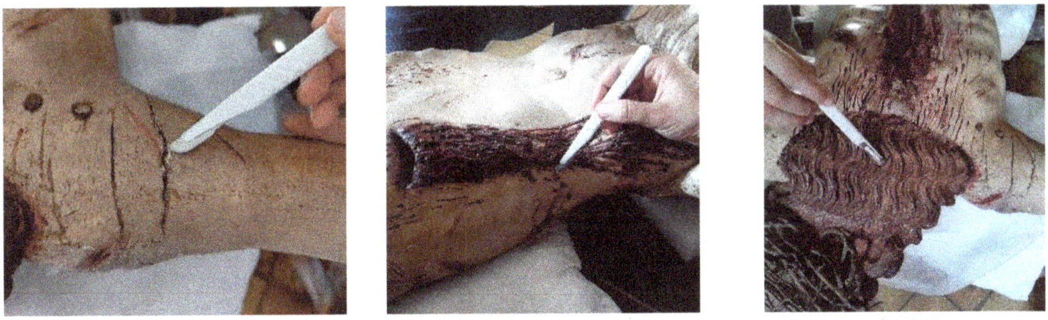

(a) CB_01 (b) CB_02 (c) CB_03

Figure 3. List of micro-fragments sampled from the Crucified Christ at Sanctuary of *Sant'Umile da Bisignano*: (**a**) CB_01, from flesh tone pictorial layer, taken along a fracture between the shoulder and the arm (left side); (**b**) CB_02, micro-fragment of paint layer (dark red) from the rib cage of the Christ at the site of blood (left side); (**c**) CB_03, wooden micro-fragment from a fracture at the back of the Christ's head.

2.2. Non-Invasive Methodologies

A portable direct digital X-ray radiography system was employed to understand the construction techniques of the three crucifixes. This diagnostic technique allows for the non-invasive analysis of the internal structures of artworks by studying the material discontinuities along the whole involved volume based on their radiopacity. This powerful investigation method provides information on the conservation state, detecting lacunae, fractures, defects, entomological attached evidence, and indications of the construction technique [16]. The portable X-ray system used for this study consists of a portable X-ray tube Poskom X+ (Northbrook, IL, USA), model PXP-100 CA, voltage from 40 to 100 kV equipped with a positioning system and an optical aiming system; a direct digital detection system: the AGFA DR 14e digital radiography plate, which consists of an amorphous silicon TFT detector, with a pixel pitch of 150 μm, active matrix 2336 × 2836 pixels with an active area of 430 mm × 350 mm, 16-bit images, and a minimum spatial resolution 3.36 lp/mm.

A preliminary cover meter survey was carried out, which aimed to correctly plan the in-situ RX acquisition and to optimise the projection of different X-ray expositions. This method allows a fast localisation of metallic structures within the wooden volume of the sculptures, measuring the absorption of the magnetic field generated by the same equipment. The metallic inserts are identified by slowly moving the probe over sensitive areas to locate the points and directions of maximum signal. The cover meter system used is the Profoscope+ by Proceq (Screening Eagle Technologies, Schwerzenbach, Switzerland).

Ultraviolet-induced visible fluorescence investigation allows for the observation and acquisition of high-resolution images of optical fluorescence stimulated by ultraviolet illumination, for the non-invasive and immediate detection of original materials and restoration materials constituting the analysed artistic surface [17–22]. The analysis also allows for the assessment of the conservation state and ageing of the protective paint, as well as the monitoring of restoration phases, for example, verifying the removal of altered paint layers. For acquisitions in optical fluorescence stimulated by UV radiation, a CCD photosensor (multispectral system Samsung NX500 28.2 MP, customised by Madatec srl, Monza, Italy) was used, equipped with a UV-IR filter to allow the detector to only reach the visible fluorescence radiation emitted by the examined materials. Visible fluorescence is induced by ultraviolet light sources (two LED sources filtered with an emission peak centred at 365 nm, positioned at 45 degrees to the observed surface).

X-ray fluorescence spectrometry (XRF) is a non-invasive analysis that allows the identification of chemical elements present in the surface materials, performed using a portable spectrometer [16,23]. In this study, this analytical technique was employed on the polychrome statue of the Crucified Christ from the Saint Salvatore Church at Gangi (Palermo), for which no sampling was conducted for the characterisation of the pictorial layers. The portable XRF spectrometer (Amptek Inc. Bedford, MA, USA) used for the multi-elemental analysis of pigments consists of an X-ray tube (Mini-X-Amptek, Amptek Inc. Bedford, MA, USA) equipped with a Rhodium (Rh) target, operating at a maximum working voltage of 40 kV and maximum current of 0.2 mA. Detection of the characteristic X-ray radiation emitted by the sample is energy-dependent (Energy Dispersive: ED-XRF) and enabled by a Silicon Drift Detector system (X-123 SDD—Amptek) with 125–140 eV FWHM at 5.9 keV Mn Kα line energy resolution (depending on the peak time and temperature), and a collimator of 1 or 2 mm. The energy detection range is from 1 keV to 40 keV, with a maximum counting rate of up to 5.6×10^5 cps. The primary beam is positioned perpendicular to the sample, while the detector is positioned at 40 degrees relative to the primary beam. Dedicated control software allows for the evaluation of measurements and acquisitions. For the analysis of the 9 selected areas, the following measurement parameters were set to ensure a good spectral signal and optimise the signal-to-noise ratio (SNR): voltage 35 kV, current 80 μA, acquisition time 50 s per area, working distance 1 cm.

Scanning electron microscopy coupled with energy dispersive X-ray analysis (SEM-EDS) was used for the morphological observation of the micro-samples collected and to perform microanalysis in terms of the elemental composition [24]. For this purpose, a Zeiss Crossbeam 350 microscope (Carl Zeiss SMT, Oberkochen, Germany), equipped with an Energy Dispersive Spectrometer (EDS), was used.

3. Results

3.1. *Crucified Christ from Santissimo Crocifisso Church at Cutro (Crotone)*

A preliminary cover meter survey revealed metallic elements only in the area between the head and the neck, corresponding to a junction line, observable at close vision (Figure 4). No other joints and nails were revealed in the remaining volume of the statue.

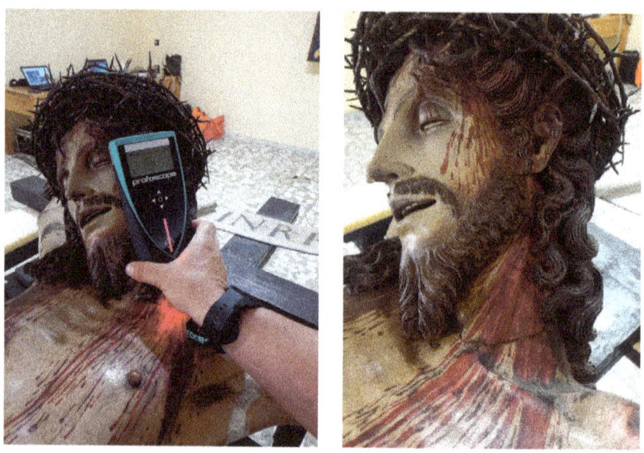

Figure 4. Preliminary survey by cover meter on the Crucified Christ of Cutro aimed to detect metal pins/nails corresponding to a junction line. The red laser indicates the exact coincidence between the sensor and the metallic elements present in the volume.

This information supported the design of successive radiographic in situ investigation in terms of the number of exposures for the area to obtain information on the direction and extension of these metal internal elements used for assembling the wooden pieces.

In situ radiographic investigations were carried out on the entire volume of the statue in an antero-posterior projection. Moreover, portions of the head, right hand, and feet were also acquired in latero-lateral or 45° oblique projection for a detailed understanding of the state of conservation in the case of the foot and hand and a more complete relief of the position and orientation of the metal anchoring pins, in the case of the head (Figure 5), starting from the information provided by the preliminary cover meter survey. Thanks to the processing and interpretation of the radiographic images, it was possible to deduce the following information for each investigated area.

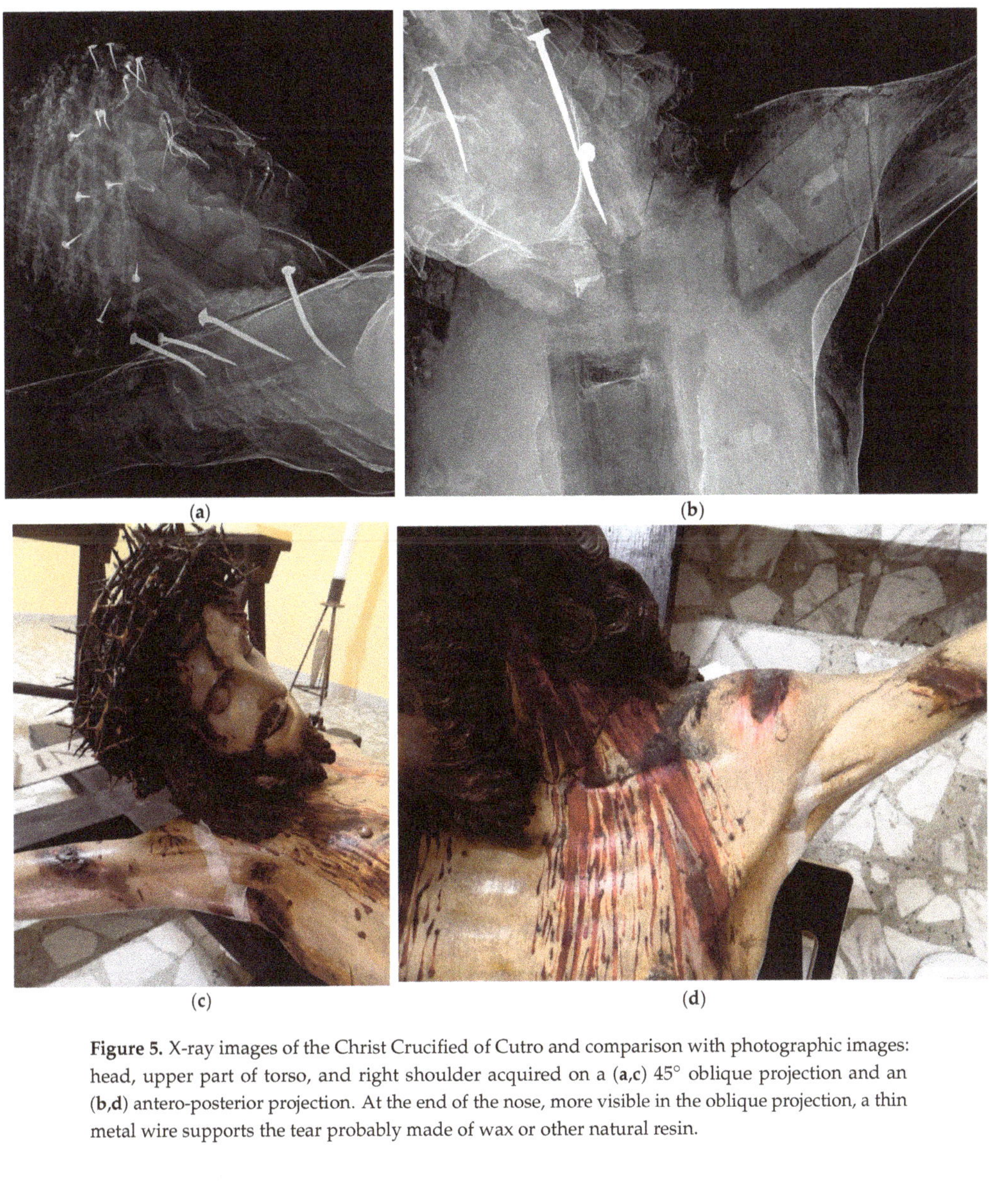

(a) (b) (c) (d)

Figure 5. X-ray images of the Christ Crucified of Cutro and comparison with photographic images: head, upper part of torso, and right shoulder acquired on a (**a**,**c**) 45° oblique projection and an (**b**,**d**) antero-posterior projection. At the end of the nose, more visible in the oblique projection, a thin metal wire supports the tear probably made of wax or other natural resin.

Concerning the head, torso, and shoulders (Figure 5a,b), the different projections carried out to understand the anchorage to the main stem showed the shape and direction of the four long nails documented at the base of the nape.

This allowed the unevenness of the surface at the back of the hair to be correctly interpreted as the areas of insertion of the metallic nails.

At the tip of the nose, a thin inserted metal wire can be seen to create a support for the applied material, forming the volume of a tear (Figure 5a). Other small metal nails were used to fix the crown of thorns on Christ's head (Figure 5a). There are no metal elements at shoulder height, but the joints between the arms and the main stem composed of tenons and wooden dowels are clearly visible. The direction of the tenon fibres, in continuity with those of the arm, is perpendicular to that of the fibres of the main stem on which it is fixed. The dimensions of the housing (mortise) are larger than those of the tenons. Between the left shoulder and the neck there is a radiopaque area, probably related to an integration area to fill a gap in the volume of the wooden stem (Figure 5b). Along the joints, irregular areas of lower radiopacity (darker areas) can be observed, which can be correlated to pictorial lacunae of the original layering. The contour lines of the different wooden pieces constituting the head and neck, which are anchored to the main stem are partially distinguishable. Moreover, two wooden dowels inserted into the chest, carved at one end to form nipples have been also documented (Figure 5b). On the abdomen and loincloth, the extension of the rectangular carving on which the spine is collocated and the presence of two little nails probably inserted to anchor it to the base have been documented (Figure 6a). The blood gushing from the Christ's side is characterised by high radiopacity, suggesting the presence of a high atomic number chemical element, a constituent of pigment in mixture (e.g., lead, mercury, etc.). At the loincloth (Figure 6b), it is possible to follow the outline of the irregularly shaped dowel applied on the back and the lacunae in the pictorial layer along the joint.

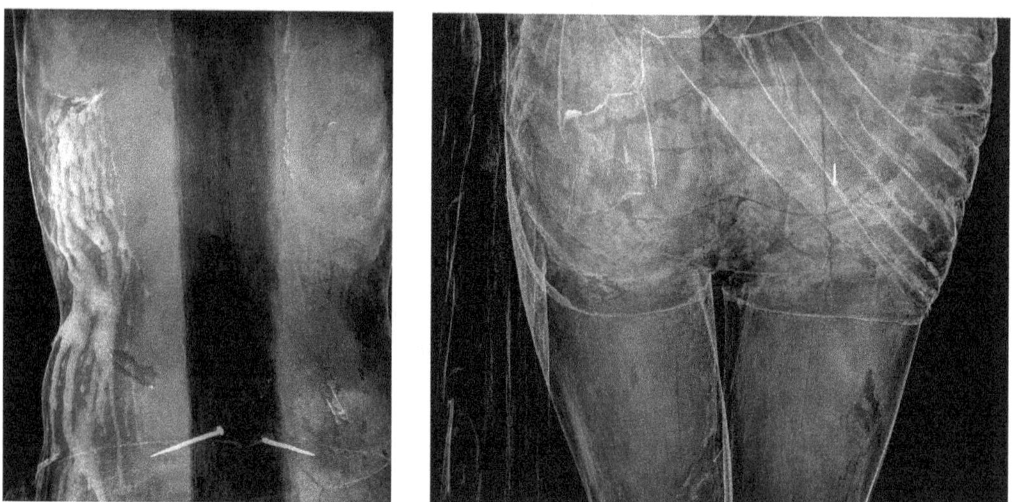

(a) (b)

Figure 6. X-ray images of the Christ Crucified of Cutro: (**a**) torso and (**b**) loincloth exposed by antero-posterior projection.

On the left side, there is a small nail that probably attaches the dowel to the main stem. Radiographic examination confirms that the loincloth was also obtained from the carving of the single trunk that makes up the body, as no joint lines, superficial fractures, metal anchors, or nails were observed in the analysed volume (Figure 6b).

The observations made on the arms–hands and legs–feet show no fractures, metal nails, or joint lines between the parts (Figure 7). This evidence confirms that the arms–hands system is carved from a single wooden piece, hooked at shoulder height. The blood drops on both hands and feet are radiolucent, suggesting a different mixture from the blood drops on the right side of torso. In general, there is little evidence of damage caused by xylophagous insects and limited lacunae and/or abrasion of the original pictorial layers, confirming the good state of conservation of the wooden support and painted surface.

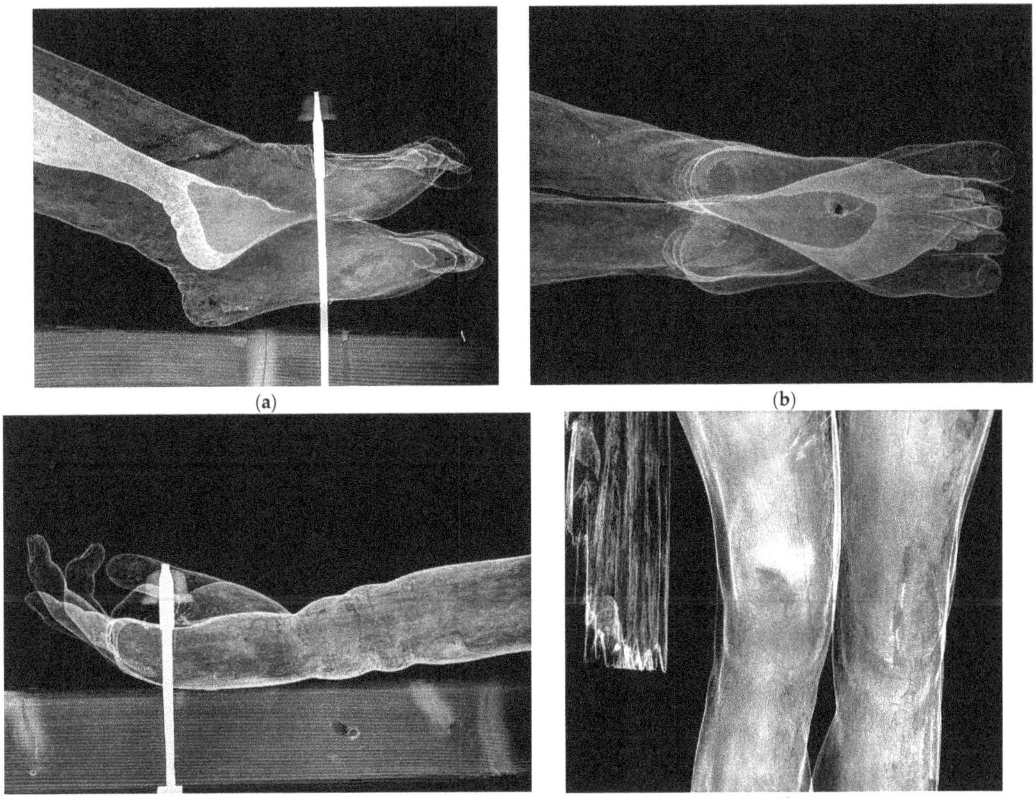

(a) (b) (c) (d)

Figure 7. X-ray images of the Christ Crucified of Cutro: (**a**,**b**) feet, (**c**) right hand, and (**d**) legs exposed by lateral (**a**,**c**) or antero-posterior projection (**b**), after modern nail removal, and (**d**).

SEM-EDS analyses performed on micro-fragments taken from the original surface of both the front and back of the Crucified Christ allowed the identification of historical pigments, excluding modern pigments due to previous restorations, localised thank to the preliminary UV fluorescence observations.

The following pigments were identified on the flesh tone layers and blood drips at the level of the rib cage: lead white (white lead carbonate, $(PbCO_3)_2 \cdot Pb(OH)_2$) [25,26], cinnabar/vermilion (HgS) [27], red lead (Pb_3O_4) or litharge (PbO) [26–28], and iron oxide based (ochres and earth pigments) [29]. In all the analysed samples, the presence of high sulphur and calcium contents was verified, attributable to calcium sulphate (gypsum), which can be attributed to the preparation layer.

The SEM-EDS spectrum acquired on CC_01 and CC_02 samples from the flesh tone layer show the presence of a first pictorial constituted by lead white and a low amount of clay mineral-based pigments (Figure 8). The overlapped dark red layer reveals the use of cinnabar/vermilion (HgS) (Figure 9). Moreover, the presence of additional chromophore elements, such as red lead or litharge, is not excluded; the latter may have also been used as a drier for protective varnishes [30]. Finally, the EDS analysis on internal layer confirms the identification of calcium sulphate as a constituent of the preparation layer (Figure 10).

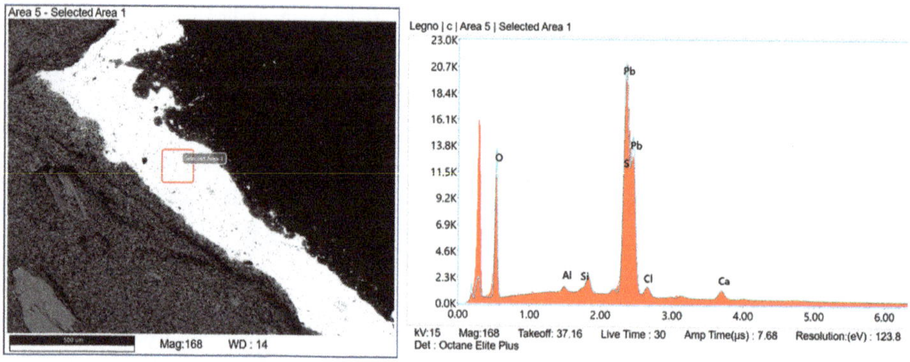

Figure 8. Christ Crucified of Cutro: SEM-EDS analyses on flesh tone pictorial layer of CC_01 sample.

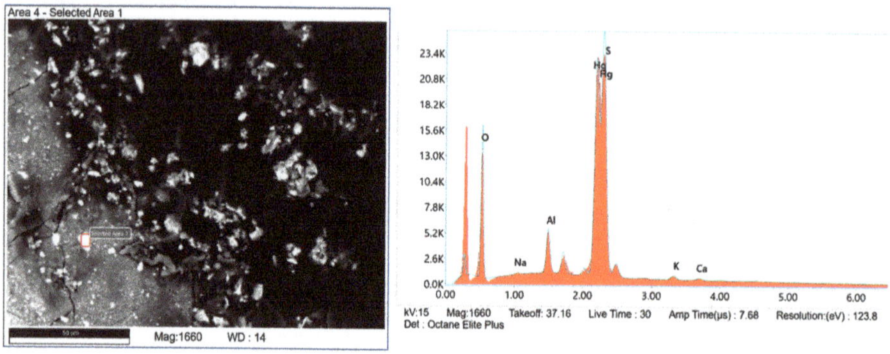

Figure 9. Christ Crucified of Cutro: SEM-EDS analyses on surface dark red pictorial layer of CC_01 sample.

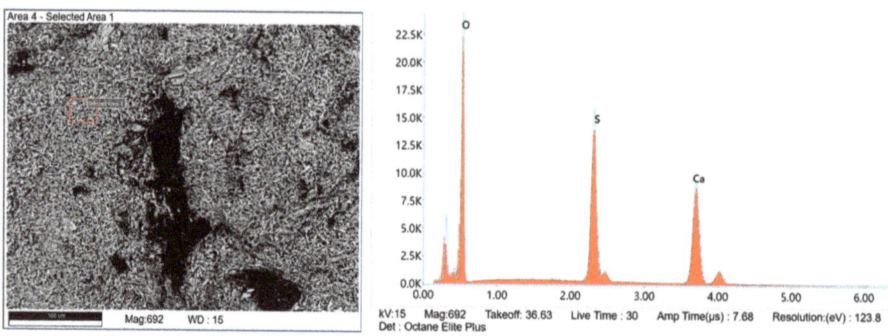

Figure 10. Christ Crucified of Cutro: SEM-EDS analyses on preparation layer of CC_01 sample.

Sample CC_03 (Figure 11), a micro-fragment from the drip of blood radiopaque in RX image, shows the presence of Hg and S, suggesting the use of cinnabar/vermilion, mixed with a lead-based pigment. Traces of phosphorus could indicate the use of an organic matrix for thickness applied to make the volume of blood drops. The presence of lead could be related to litharge (lead oxide) added as siccative [30]. Furthermore, the presence of aluminium, magnesium, and potassium suggest the use of mordants for the red lake, in a mixture to obtain the dark intense colour tone [31].

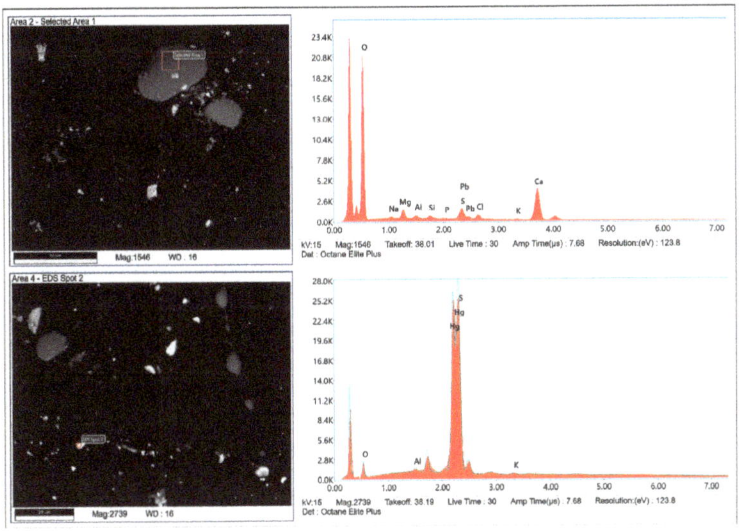

Figure 11. Christ Crucified of Cutro: SEM-EDS analyses on CC_03 sample from thick blood layer.

For the identification of the wood species, sample CC_04 taken from the top of the head was analysed by SEM-EDS (Figure 12). The SEM images showed the presence of a surface homogeneous layer that, from the EDS analysis, related to calcium sulphate constituting the preparation layer [32]. The morphological evidence provided by SEM observations of the wooden fragment suggests the use of poplar (*Populus*).

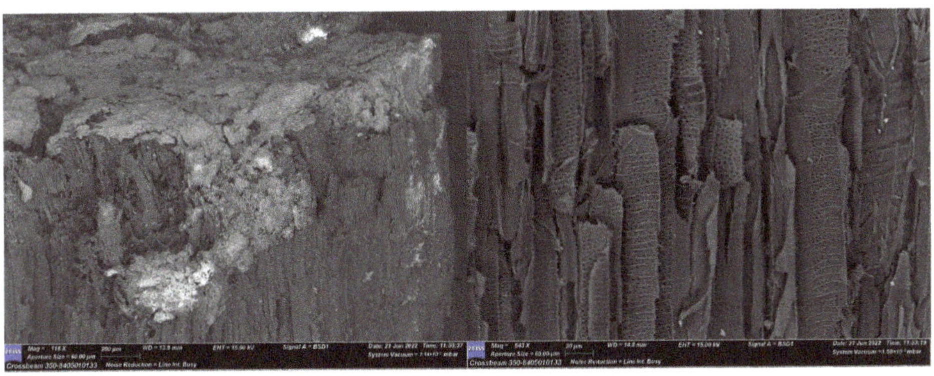

Figure 12. Christ Crucified of Cutro: SEM images of CC_04 wooden sample and preparation layer.

The genus Populus does not allow the precise identification of different species belonging to the group. The technological features of the wood are of three more diffused and similar species (*P. alba*, *P. nigra*, *P. tremula*), and there is a lack of knowledge of the correct species [4–6]. However, no information on the wood used for the body of Christ (which, as observed in the X-ray, appears to consist of a single central stem to which the head and arms were anchored) was obtained, due to the inability for further sampling.

3.2. Crucified Christ, Sanctuary of Sant'Umile at Bisignano (Cosenza, Calabria)

In this case, the preliminary cover meter survey also revealed metallic elements in the area between the head and neck and at the end of the torso. The low radiopacity of the pictorial and preparation layers allowed documentation of the features of the wooden volume of the whole sculpture, analysed by antero-posterior projection. In addition, portions of the head and neck were also acquired by latero-lateral and oblique 45° projection for a more complete understanding of the dimension and orientation of the nails already localised by the first cover meter survey.

The radiography images of the head and the upper part of torso (Figure 13) revealed the shape and direction of the three nails documented at the base of the nape of the neck; moreover, small nails applied to the head used to fix the crown of thorns were also documented. At shoulder height, there are no nails or metallic elements, but the assembly of the arms to the main stem by means of tenons and pegs (or thorns) is clearly visible. In the central part of the torso, the perimeter of the hollow rectangular carving is documented (Figure 13b). The absence of pictorial layers on the inner surface of the carving and the reduced thickness of the pictorial layer make the course of the fibres of the wooden stem used to carve the body legible.

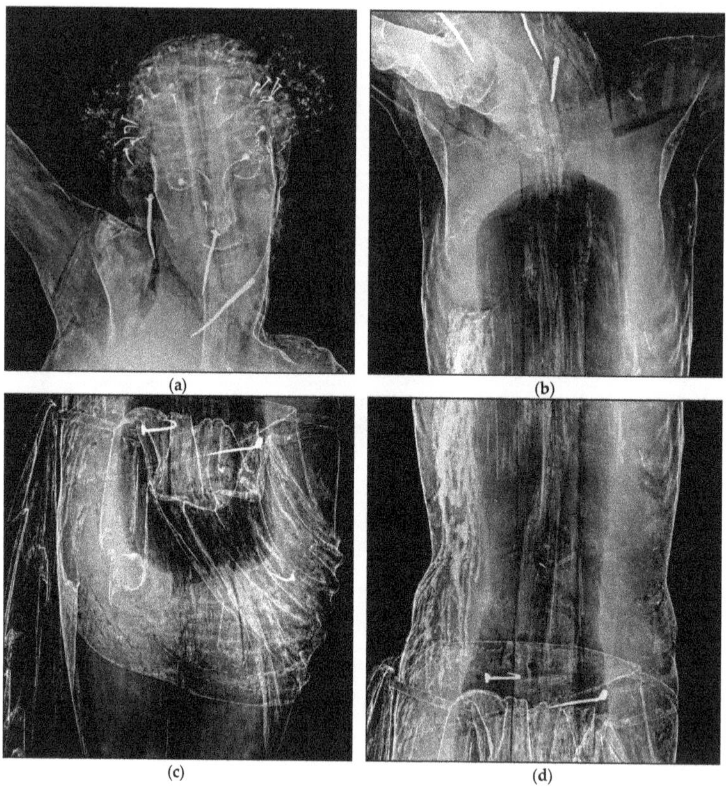

(a) (b) (c) (d)

Figure 13. X-ray images of Christ Crucified from Bisignano: (**a**) head and part of the right shoulder (45° oblique projection); (**b**) neck and chest; (**c**) loincloth and part of legs; (**d**) abdomen and upper portion of the loincloth (antero-posterior projection).

The direction of the fibres of the head suggests that the piece of wood used comes from the same trunk, this one carved and shaped to obtain the desired inclination. The head and the rest of the main trunk return the same level of radiopacity, and we note the greater density of fibres in continuity in the central part of the two assembled parts (Figure 13a,b). There is also evidence of two wooden thorns inserted into the chest and carved at one end forming nipples, very similar to those documented in the Crucified Christ of Cutro. The blood flowing from Christ's chest (Figure 13c) is characterised by high radiopacity, suggesting the use of one or more pigments with a high atomic number such as Pb or Hg. At the end of the torso (Figure 13c,d), the X-ray images show the presence of two nails probably inserted to anchor a rectangular wooden element, applied to close the cavity carved on the back of trunk.

The loincloth (Figure 13d), as well as the legs and feet (Figure 14a,b) appear to have been carved from a single trunk constituting the body: neither joint lines, anchors, nor nails were observed. As for the hands and wrists (Figure 14c), no metal nails or pins were found, and the low radiopacity of the pictorial layer allowed us to confirm the absence of junction systems. The RX investigation highlights that the arms–hands parts are each carved from a single wooden piece attached to the torso at shoulder height. The blood drops on both the hands and the feet (Figure 14b,c) are characterised by the same radiopacity found in the drops on the chest (Figure 13). In general, there is little evidence of xylophagous insect damage and limited lacunae and/or abrasion of the original pictorial layers. In contrast, the wooden element constituting the closure of the hollow part at the back appears more degraded, showing a higher density of holes due to the past attachment of xylophagous insects and unevenness in the wooden thickness involved in RX transmission.

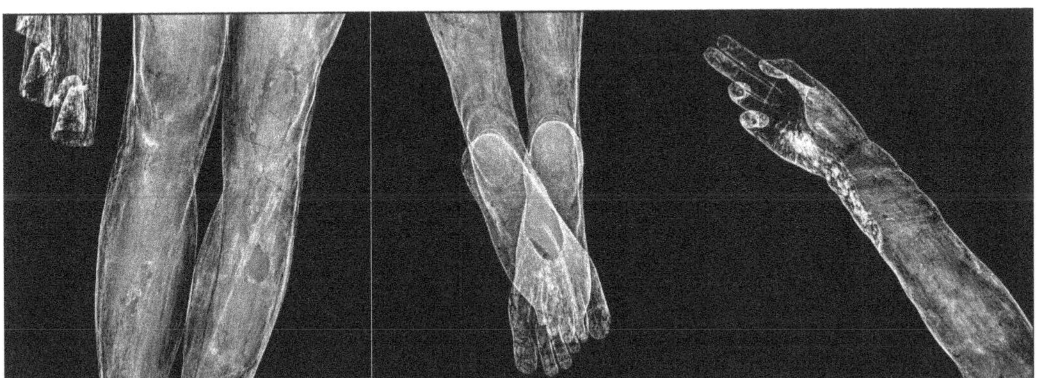

(a) (b) (c)

Figure 14. X-ray images of Christ Crucified from Bisignano: (**a**) the knees and legs; (**b**) the feet–ankles; (**c**) the right arm–hand.

Preliminary observation of sample CB_01 (from flesh tone layer) under a digital optical microscope (Figure 15) revealed the presence of a white preparatory layer on which a pinkish-coloured pictorial layer, characterised by a darker pink thin film, has been overlapped. Laterally on the fragment, fabric fibres were observed that could be traced to the use of a canvas, applied in some areas of the sculptures, confirming the evidence observed through X-ray images at high magnification.

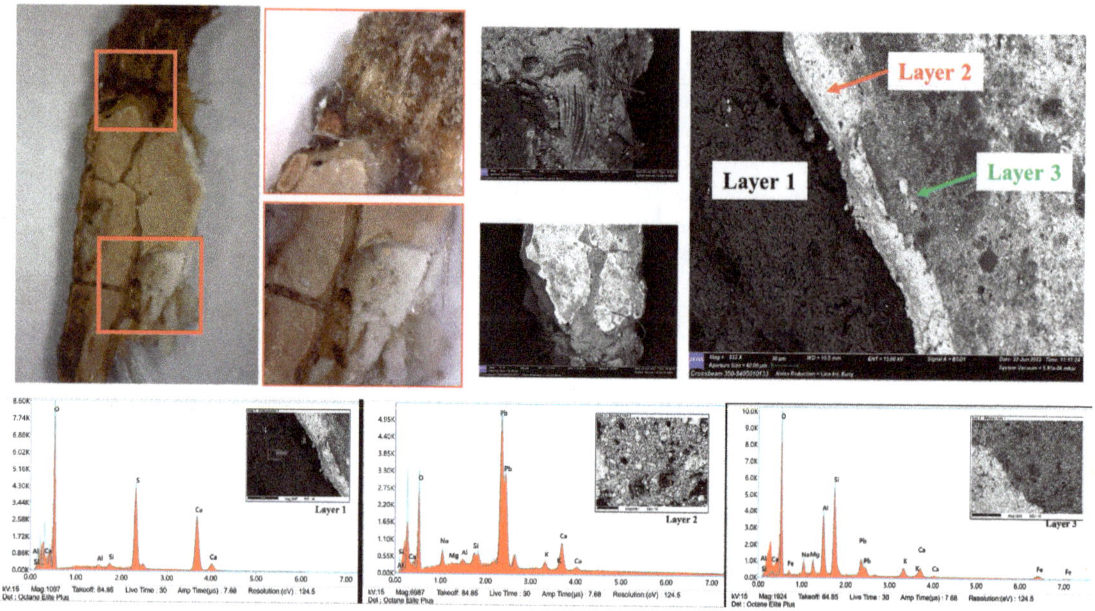

Figure 15. SEM analyses and digital light microscope observations of samples CB_01. EDX spectra obtained from each layer: preparation (Layer 1); flesh tone pictorial layer (Layer 2); darker pink thin film (Layer 3).

The described stratigraphy and the presence of canvas filaments were also documented through SEM image and EDS analyses (Figure 16). The EDS spectra acquired on each three layers document the following stratigraphic sequence, from the bottom: a preparation layer (layer 1), consisting of calcium and sulphur (calcium sulphate) [24,29,31]; a middle layer (layer 2), predominantly composed by Pb, attributable to lead white (lead carbonate, $(PbCO_3)_2 \cdot Pb(OH)_2$) [26]; and a surface layer (layer 3), applied to make the pinkish layering tone, consisting of iron oxide pigments, clay minerals, and probably red lacquer, suggested by the presence of potassium (a chemical element that could be related to the mordant for the red lake) [12,31,32].

Preliminary observations by using a digital optical microscope and SEM of CB_02 sample document that the micro-fragment (from a drop of blood in the ribcage) consists of a homogenous matrix (Figure 16). In particular, the SEM image reveals the presence of radiopaque grains of pigment, and the EDS spectra acquired on these irregular grains detected mercury and sulphur, attributable to the use of cinnabar/vermilion [33,34] and aluminium and sulphur, which can be correlated to a possible use of a mordant in the production of red lake [35,36]. These chromophores (cinnabar/vermilion and red lake) are immersed in an organic matrix that was probably used to three-dimensionally mould the blood drops [37,38].

The xylotomic investigation for identifying the wood species [39] was performed on sample BC_03, from the hair corresponding to a lacuna). The optical microscope observation was carried out in three cross sections: transverse, tangential longitudinal, and radial longitudinal (Figure 17).

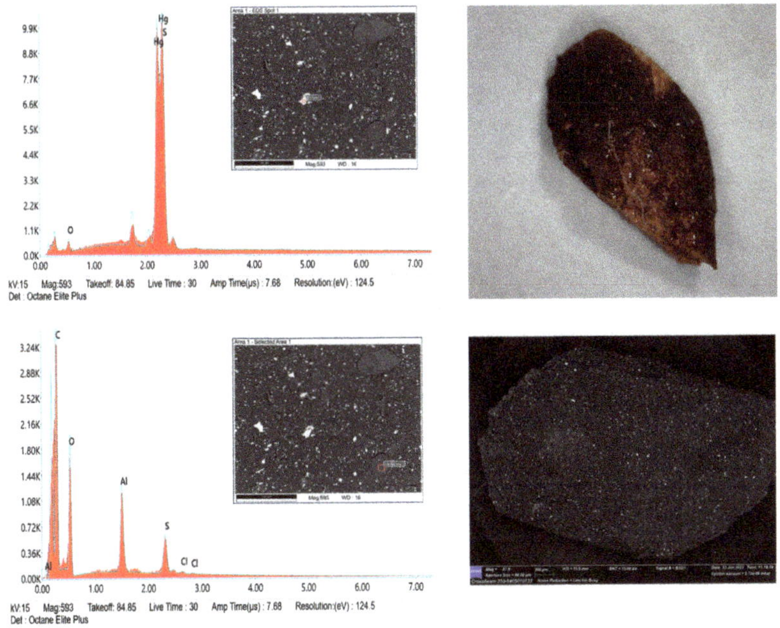

Figure 16. Digital optical microscope observations and SEM images of samples CB_02. EDX spectra obtained on radiopaque pigment and radiolucent dyes grains, added in the organic matrix.

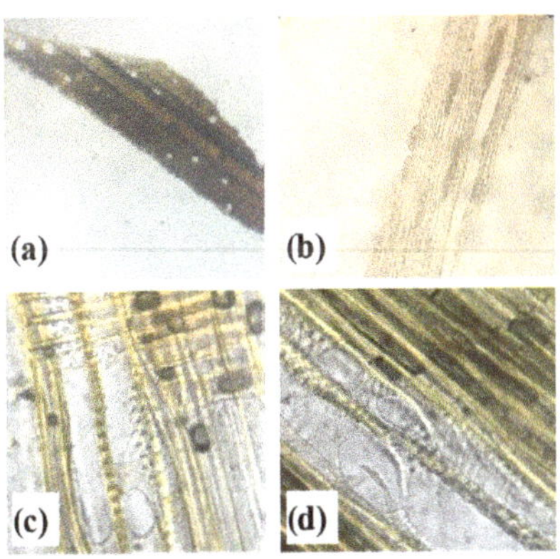

Sample CB_03	Wood of *Acer sp.*
Transverse section	Incomplete ring of diffuse porous hardwood with widely spaced round solitary vessels or in radial rows of 2-4 elements
Tangential section	Homogeneous parenchyma rays 1-4 cell axes wide
Radial section	Homogeneous radii, simple vessel perforation and vessels with spiral thickening

Figure 17. Light microscope images of three cross sections obtained from sample CB_03 and characteristic of the identified wood genus: (**a**) transverse; (**b**) tangential; (**c**,**d**) radial sections. Dimension of observed area is 50 µm.

Schweingruber's keys [40,41] were used for identification: based on the observed typical feature set, the sample is attributable to maple wood (*Acer* sp.), a genus represented by several species that are difficult to distinguish microscopically, especially in the case of small fragments. Maple is a suitable wood for carving, although its use in ancient wooden sculpture is poorly documented [42,43].

3.3. Crucified Christ from the Church of San Salvatore from Gangi (Palermo)

The preliminary inspection by cover meter indicated the presence of metallic nails or pins in the right knee, loincloth knot, head, lower part/neck, and abdomen. Radiographic investigations aimed at understanding the construction technique had limitations due to the high radiopacity of the pictorial materials or the overlapping of more pictorial layers that homogeneously covered the entire wooden surface.

Despite this restraint, information on the interconnection systems between the different carved wooden portions was deduced. The RX projections performed for the head and right shoulder (Figure 18a,b) showed how the anchorage of the head to the main trunk was made by three long nails documented at the base of the nape (Figure 18a,b). A more radiopaque straight line at the base of the head is observed on radiography. The areas of high radiopacity present on the upper portion of the head, particularly along the right side, can be interpreted as fillings (Figure 18c).

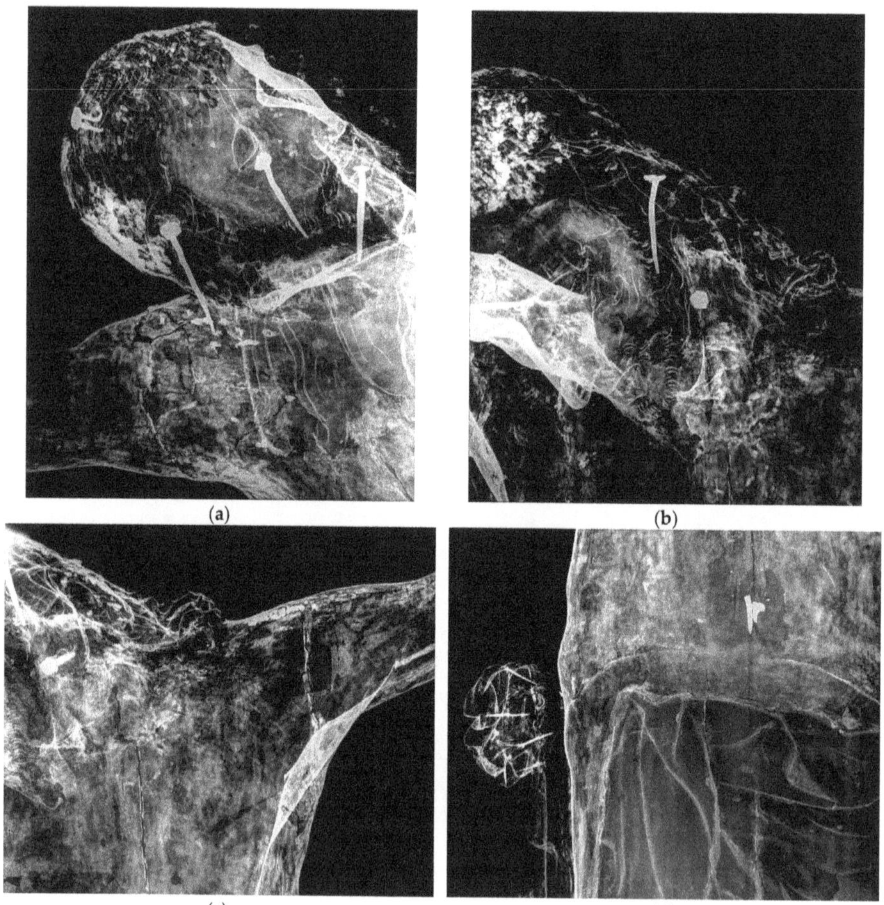

(a) (b) (c) (d)

Figure 18. X-ray images of Christ Crucified from Gangi: (**a**,**b**) head and part of the right shoulder; (**c**) neck and left shoulder; (**d**) loincloth.

At the level of the shoulders (Figure 18c), no metallic elements are found. The high radiopacity of the surface pictorial layers does not allow one to clearly observe the wooden volume, but the pattern of the surface cracks in the pictorial surface allowed us to hypothesise a mortise–tenons system, fixed by means of dowels (or pins), for the junctions of the arms to the main trunk. There is a longitudinal–radial shrinkage crack, running from the neck to the navel, that confirms the use of a single trunk to carve the body of the Christ, excluding the head and arms.

The lower part of the torso is not directly visible due to the high radiopacity of the pictorial layer. Instead, the thin surface fracture lines (Figure 18d) affect the pictorial layers and, finally, the lacunae in the underlying pictorial layer and allow documentation of the past conservation state of the sculpture. It seems highly probable that the longitudinal crack at the level of the navel (Figure 18d) was already present in the past and partially reinforced with the two small nails.

In correspondence with the loincloth, a greater radiolucency can be observed (Figure 18d) due to the absence of the lead white pictorial layer. In this portion, thanks to the low radiopacity, the good state of preservation of the wooden support and the absence of degradation attributable to xylophagous insects were confirmed.

The different response in X-ray of the right side of the loincloth and the high number of nails observed at the knot, (Figure 18d), which do not coincide with the other cases, confirms that this portion was affected by the remaking of the entire wooden volume and not only by a superficial pictorial integration. The two small aligned metal elements revealed in the right knee (Figure 19a) are interpreted as anchoring pins for the dowel that was used for the volume construction. The radiographic investigation of the feet and ankles (Figure 19b) verified the absence of metal nails or pins, and the high radiopacity of the pictorial layer (whose lacunae and fractures are well-followed) does not allow us to visualise the perimeter of the junction system used for the attachment of the feet to the legs. Similar observations were carried out for the hands and wrists of this Christ (Figure 19c); only a thin metal wire used to anchor the fragment of a finger can be observed. It worth noting that the thick drops of blood appear radiopaque, deducing that cinnabar or lead-based mixtures were not added to the organic matrix, as revealed in the Cutro and the Bisignano Crucifixes.

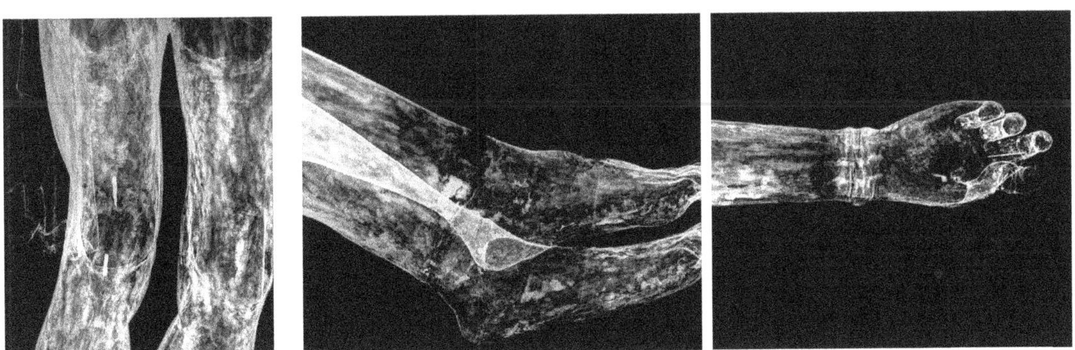

(a) (b) (c)

Figure 19. X-ray images of Christ Crucified from Bisignano: (**a**) the knees and legs; (**b**) the feet–ankles; (**c**) the left arm–hand.

For an immediate and preliminary inspection of the pictorial surface and subsequent assessment of the areas affected by retouching or pictorial integration, an observation of the entire artwork was carried out under 365 nm UV LED. This investigation was also fundamental for assessing the areas to be investigated under X-ray fluorescence to study the historical pictorial layers overlapped over time. Observation under UV light provided an initial mapping and differentiation of the protective or consolidating materials applied to the surfaces (Figure 20).

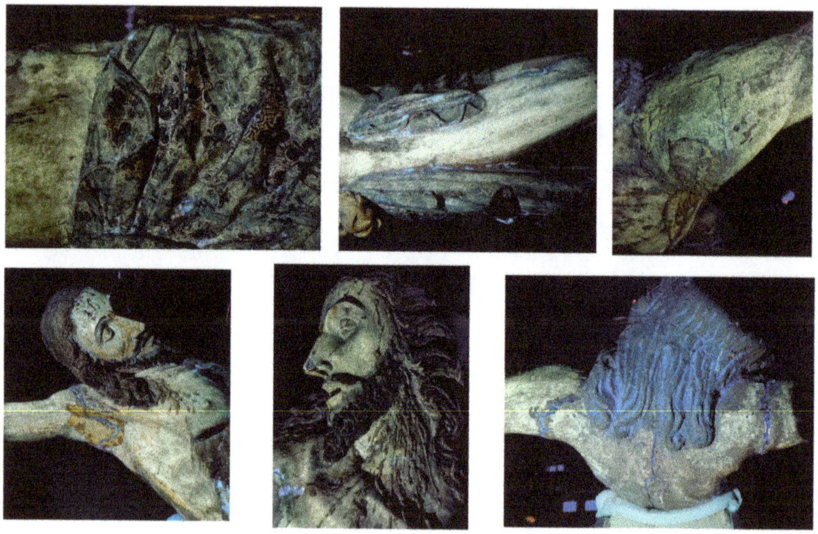

Figure 20. Visible fluorescence under UV LED observation of the pictorial surfaces of Christ Crucified of Gangi, before the restoration works.

On the right side of the body and on the lateral portion of the loincloth, the inhomogeneity of the visible fluorescence behaviour can be attributed to residues of the previous aged protective varnish that had not been completely removed [18,19]. The gilding layers of loincloth with floral decoration showed no evidence of repainting, but the unevenness found is due to the presence of an unevenly applied protective coating that can be localised by its bluish fluorescence response [20]. In some areas, a yellow–orange fluorescence indicating the presence of an organic material (probably shellac) [17], which appears more unevenly on the face and torso of Christ, was observed.

Discontinuity in the superficial pictorial layer was more clearly identified due to the different response under UV LED of the darker underlying layer of the flesh tone [21]. The beard was also affected by repainting; in fact, a darker colouring was observed along the edge that was not affected by the older layer of varnish, distinguishable by the yellowish fluorescence [22]. On the back, in particular, the absence of fluorescence of the repainting is evident, which highlights the path of the fractures generated at the joints in the shoulders and the longitudinal crack of the central part of the torso (also observed in X-ray) [18].

Finally, four surface areas were analysed by an X-ray fluorescence spectrometer to identify the pictorial layers' composition. The chemical analysis aimed to preliminarily qualitatively characterise the complex superimposition of the pictorial layers applied over time, observed by means of a digital optical microscope in correspondence with falls or lacunas of upper layers (Figure 21).

Above the preparation, at least three layers were identified (Figure 21): (1) the first constitutes a probable imprimitura; (2,3) the other two layers are related to layers of flesh tone juxtaposed over time to adapt to the artistic and cultural style typical of *Fra' Umile* (repainting, dated back to 1656) and to compensate for extensive colour losses in the first layer, as revealed by the RX images. The XRF analyses performed on the three selected areas in correspondence with the pictorial layers identified the presence of historically used pigments and excluded the presence of modern synthetic pigments.

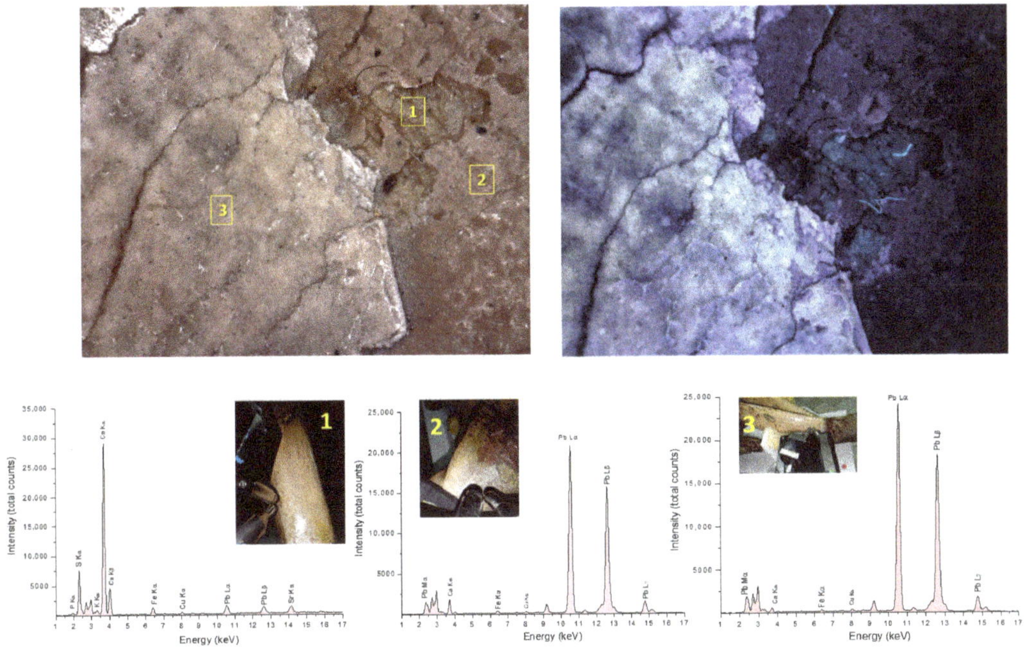

Figure 21. Vis and UV optical microscopic details of the performed XRF analysis with corresponding XRF spectra of the identified layers and pigments, from left: 1, preparation/imprimitura; 2. original pictorial layer; 3. historical repainting layer (1656).

Lead white and iron oxide pigments (ochres or earths) have been identified both in the underlying layers and in the more superficial pictorial layers. In the red layer of the oldest pictorial layer, localised at a lacuna of the surface layer of the foot, the use of lead white and ochres or earths and a low amount of cinnabar/vermilion were identified (Figure 22). In this case, the xylotomic investigation was not carried out, as no samples were authorised.

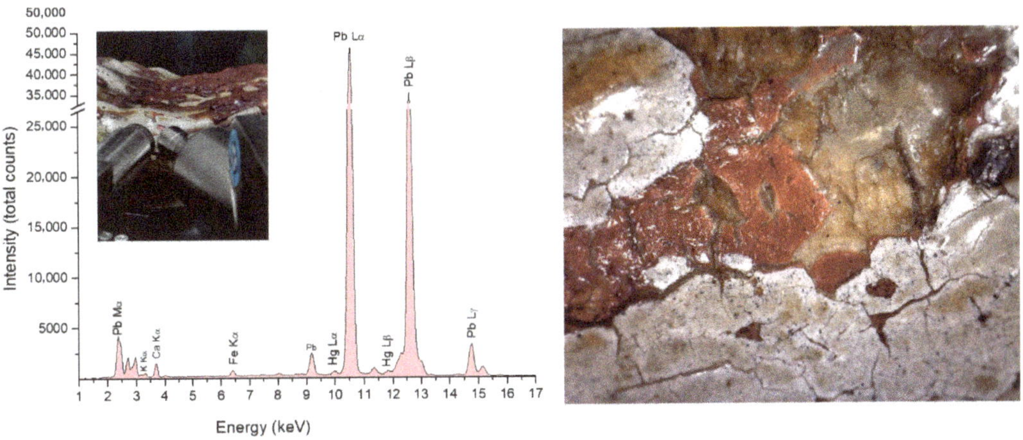

Figure 22. Red pigments localised on the oldest pictorial layer, for which the use of red lead-based pigment has been suggested by high XRF signal intensity of lead.

4. Discussion

All three Crucifix sculptures have been involved in a systematic RX diagnostic imaging investigation. Although the artworks are similar, some technical and conservative differences have been clearly documented. Firstly, with regard to the presence of metal elements, the Christ Crucified in Cutro and the one in Bisignano show a concentration of metallic elements mainly between the head and neck. In the Crucifix in Gangi, in addition to the presence of metal elements such as nails between the head and neck, metal elements were also observed on the right knee, abdomen, and loincloth knot, the latter applied during a previous restoration. From the point of view of the characteristics revealed by the X-ray images, distinctive details emerge in each sculpture. In the Crucifix from Cutro, for example, one can clearly see the nails used to fix the head to the bust and small nails used to fasten the crown of thorns; in addition, the anchoring of the arms was carried out without metal joints but with a tenons–mortise system and pins. The entire sculpture, with the exception of the head and arms, was carved by using a single wooden trunk. These construction characteristics can also be found in the Crucifix in Bisignano. In both cases, Cutro and Bisignano, the rectangular cavity present in the central part of the torso along the entire length was revealed. This evidence agrees with what is reported in the literature [4–6]. Most of the statues, indeed, are made of a single stem, excavated on the back of the statue in order to prevent the opening of the shrinkage cracks, and all the components of the figure that cannot be included on the projection of the stem are carved on external wooden parts, jointed with mechanical connections, strengthened by organic glue.

The presence of a canvas (an incamottatura to reduce the movement of the wood along the joints) along the left arm of Crucifix in Bisignano, constitutes a different technical choice from that of Cutro.

In the Crucifix in Gangi, radiographic investigations aimed at understanding the construction technique showed limitations due to the high radiopacity of the double lead-based pictorial layers covering the entire wooden surface. However, some distinctive details emerged, such as the radiopaque stucco integration on the head, face, and loincloth, indicating probable restorations over time after its creation.

For this Crucifix, the anchoring system of the arms via the tenon–mortise system and of the reclining head carved separately and fixed to the trunk via nails remain the same as the two previous cases. However, the rectangular cavity in the trunk was not revealed in the Gangi Crucifix, and the absence of this technical solution in fact caused the formation of shrinkage fractures, highlighted on the x-ray images.

From a conservation point of view, all three artworks show a good general conservation state, with little evidence of xylophagous insect damage and limited lacunae or abrasions of the pictorial layers. Thanks to the wooden samples analyses, in two cases, it was possible to identify the type of wood used in the production of the sculpture or part of this. For the Cutro Crucifix, the use of poplar (*Populus*) was identified by morphological SEM images, while for the Bisignano Crucifix, the use of maple wood (*Acer* sp.) was identified by cross-sectional optical microscope observation. The former appears more commonly used than the latter in the 16th and 17th century wood sculpture production. It should be noted that the identification of the wood species relates to the head of the crucifixes investigated; the impossibility of systematic sampling does not allow for a complete identification of any different genus used for carving the arms and the main trunk.

In the case of Bisignano, however, it was hypothesised, on the basis of the continuous trend and the same density of the fibres observed in RX, that the head was obtained from part of the same trunk.

In order to fully understand the differences and similarities of the painting techniques used in the three crucifixes, SEM-EDS investigation and X-ray fluorescence analyses, where sampling was not permitted, were carried out. The SEM-EDS analyses performed on the Cutro Crucifix allowed us to identify the use of traditional historical pigments, such as lead white, cinnabar/vermilion, and minium or litharge. These pigments were found in the flesh tones and blood drippings. Similarly, analyses carried out on the Crucified Christ in Bisignano confirmed the use of these pigments, and iron oxide pigments have also been hypothesised [31,44]. Both crucifixes show a similar stratigraphic sequence, with a preparation layer consisting of calcium sulphate (gypsum) [29]. Analyses carried out on the Crucified Christ in Gangi by X-ray fluorescence investigation also revealed the following pigments: lead white and iron oxide pigments were identified for the flesh tone layers. For all three artworks, the use of cinnabar/vermilion mixed with minium or litharge with red lake in the organic matrix appears likely in the blood areas [17]. Table 1 shows the main identified chemical elements found in the three works with the identification hypothesis of the preparation layer and the probable pigments used for the embodiment and blood drops.

Table 1. Main results obtained through elemental analysis on the three crucifixes analysed.

Artwork	Sample	Colour	Identified Elements	Preparation Layer	Identified Pigments
Crucified Christ from Santissimo Crocifisso Church at Cutro (Calabria)	CC_01	Flesh tone	Ca, S, Hg, Si	Calcium sulphate	cinnabar/vermilion
	CC_02	Flesh tone	Ca, S, Pb, Hg, Si, Al	Calcium sulphate and lead white	cinnabar/vermilion; earth; lead white; litharge
	CC_03	Blood Dark red	Ca, S, Pb, Hg, P, Si, Al, Mg, K	Calcium sulphate	cinnabar/vermilion, earth; lead white; minium; shellac; litharge
Crucified Christ from Sanctuary of *Sant'Umile da Bisignano* (Calabria)	CB_01	Flesh tone	Ca, S, Pb, Si, Al, K, Mg, Fe	Calcium sulphate and lead white	cinnabar/vermilion; earth; lead white; minium; shellac; litharge; pigment
	CB_02	Blood Dark red	Hg, S; Al, Cl		cinnabar/vermilion; shellac
Crucified Christ from Saint Salvatore Church at Gangi (Sicily)	3 points	Flesh tone	Pb, Fe, Ca	Calcium sulphate	lead white; ochre or earth pigments
	1 point	Blood Dark red	Pb, S, Hg, Fe, Ca	-	red lead; low amount of cinnabar/vermilion; ochre or earth pigments

The similarities observed both in the polychrome (Table 1) and construction technique of the wooden sculptures reflect an artistic and historical tradition typical of artworks created in the period between the 16th and 17th centuries. Moreover, the observed similarities support the attribution to the same artist or workshop. In fact, although some small differences have been observed between the Calabrian works, whose attribution is known, and the Sicilian one, whose attribution is uncertain, we can affirm that they are similar in terms of the construction technique in terms of the number of pieces and the assembly methods of the wooden statue and the composition and stratigraphy of the pictorial layers

5. Conclusions

The three wooden sculptures analysed, the Crucified Christ from *Santissimo Crocifisso* Church at Cutro (Calabria), the Crucified Christ from the Sanctuary of *Sant'Umile da Bisignano* (Calabria), and the Crucified Christ from Saint Salvatore Church at Gangi (Sicilia), revealed interesting similarities thanks to the analyses conducted using a digital direct radiography imaging, SEM-EDX, and X-ray fluorescence analyses. The investigations showed similar construction techniques and pictorial materials for the three sculptures, supporting the hypothesis of a chronological coincidence and attribution to the same workshop based on stylistic analyses and historical archival documents. The similarities in construction and artistic techniques reflect an artistic tradition of the 16th and 17th centuries. These observations support the attribution of the artworks to *Fra' Umile da Petralia*, and regarding that of Gangi, to his workshop and successors relating to the pictorial materials and artistic technique used for the stylistic adaptation entrusted to *Fra Umile's* successors testified to in an archive document from 1656.

This preliminary study highlights the importance of systematic research into the materials and construction solutions of wooden crucifixes in Southern Italy, not yet widely available to the scientific community, which could contribute to the correct identification and differentiation of the active artists and their workshops. Starting from the study of works of art of certain attribution, as in this case, the systematic collection of technical knowledge on the pictorial technique, the carving and construction techniques, the assembling, the woods selected for the different parts of the statue, and those used in the different geographical areas, would enable a set of knowledge and data to support attributions even on scientific bases and not only on artistic–stylistic criteria often influenced by the difficult reading of the original models and surfaces altered by numerous overlaps and restorations.

Author Contributions: Conceptualisation, M.F.L.R., M.F.A., E.G., A.A., G.M. and S.S.; methodology, L.R., M.R., S.S. and S.A.R.; formal analysis, V.C. and S.S.; investigation, L.R., M.R., M.F.A. and S.S.; resources, M.F.L.R.; data curation, L.R., M.R., M.F.A. and S.S.; writing—original draft preparation, V.C., S.A.R. and M.F.A.; writing—review and editing, L.R., M.R. and S.S.; supervision, M.F.L.R. All authors have read and agreed to the published version of the manuscript.

Funding: This research received no external funding.

Data Availability Statement: All data are included in this paper.

Conflicts of Interest: Authors Maria Francesca Alberghina and Salvatore Schiavone were employed by the company S.T. Art-Test di S. Schiavone & C Sas, Italy. The remaining authors declare that the research was conducted in the absence of any commercial or financial relationships that could be construed as a potential conflict of interest.

References

1. La Mattina, R.; Dell'Utri, F. *Frate Umile da Petralia. L'arte e il Misticismo*; Lussografica Editoria: Caltanissetta, Italy, 1987.
2. Fazio, G.; Brancati, S. *I Crocifissi di Frate Umile e Frate Innocenzo da Petralia*; EAN: 9788894001921; Arti Grafiche Martorina: Ispica, Italy, 2019.
3. Abbate, G.; Fazio, G. *Frate Umile a Petralia Soprana. Il Suo Tempo i Suoi Luoghi*; Regione Siciliana, Assessorato dei Beni Culturali e dell'Identità Siciliana, Dipartimento dei Beni Culturali e dell'Identità Siciliana: Palermo, Italy, 2015.
4. Fidanza, G. Caratteristiche tecnologiche e formali delle specie legnose: Una verifica su statue e intagli di età moderna. In *Statue di Legno. Caratteristiche Tecnologiche e Formali delle Specie Legnose*; Istituto Poligrafico e Zecca dello Stato: Rome, Italy, 2008; pp. 33–57. (In Italian)
5. Macchioni, N.; Fachechi, G.M.; Lazzeri, S.; Sozzi, L. Timber species and provenances of wooden sculptures. Information from the collections of the National Museum of "Palazzo di Venezia" in Rome. *J. Cult. Herit.* **2015**, *16*, 57–64. [CrossRef]
6. Macchioni, N.; Lazzeri, S.; Sozzi, L.; Vitiello, R. Wooden sculptures from XVII and XVIII centuries in the region of Asti (Italy): Scientific identification of the species. *Int. J. Conserv. Sci.* **2011**, *2*, 251–260.
7. Lo Monaco, A.; Agresti, G.; Serusi, G.; Taddei, A.R.; Pelosi, C. History and Techniques of a Polychrome Wooden Statue, How an Integrated Approach Contributes to Resolving Iconographic Inconsistencies. *Heritage* **2022**, *5*, 2488–2503. [CrossRef]
8. Di Lodovico, D. Revising Devotion: The Role of Wooden Sculptures in Affecting Painting and Devotion in the Late Medieval Period in Italy (XII–XV Century). Ph.D. Thesis, University of Washington, Seattle, WA, USA, 2016; 394p.
9. Ruffolo, S.A.; La Russa, M.F.; Barca, D.; Casoli, A.; Comite, V.; Nava, G.; Crisci, G.M.; De Francesco, A.M.; Miriello, D. Mineralogical, petrographic and chemical analyses for the study of the canvas "Cristo alla Colonna" from Cosenza, Italy: A case study. *Period. Mineral.* **2010**, *79*, 71–79. [CrossRef]
10. Vizárová, K.; Reháková, M.; Kirschnerová, S.; Peller, A.; Šimon, P.; Mikulášik, R. Stability studies of materials applied in the restoration of a baroque oil painting. *J. Cult. Herit.* **2011**, *12*, 190–195. [CrossRef]
11. Miliani, C.; Rosi, F.; Brunetti, B.G.; Sgamellotti, A. Non-invasive analysis of paintings: The application of visible, ultraviolet and infrared spectroscopic techniques. *Herit. Sci.* **2013**, *1*, 8. [CrossRef]
12. Ricca, M.; Paladini, G.; Rovella, N.; Ruffolo, S.A.; Randazzo, L.; Crupi, V.; Fazio, B.; Majolino, D.; Venuti, V.; Galli, G.; et al. Archaeometric characterisation of decorated pottery from the archaeological site of Villa dei Quintili (Rome, Italy): Preliminary study. *Geosciences* **2019**, *9*, 172. [CrossRef]

13. Fermo, P.; Mearini, A.; Bonomi, R.; Arrighetti, E.; Comite, V. An integrated analytical approach for the characterization of repainted wooden statues dated to the fifteenth century. *Microchem. J.* **2020**, *157*, 105072. [CrossRef]
14. Oliveira, R.; de Paula, A.; Gonçalves, F.; Sanches, F.; Nardes, R.; Santos, R.; Azeredo, S.; Araújo, O.; Machado, A.; Anjos, M.; et al. Analysis of a wooden statue by non-destructive X-ray techniques. *X-Ray Spectrom.* **2023**, *52*, 312–322. [CrossRef]
15. Morigi, G.; Morigi, M.; Bettuzzi, M.; Casali, L.; Di Cocco, V.; Brancaccio, M. X-Ray Tomography of Archaeological Cultural Heritage Materials. In Proceedings of the LuBec 2018, Lucca, Italy, 4–5 October 2018. Available online: https://www.lubec.it (accessed on 10 December 2024).
16. Janssens, K. X-Ray Based Methods of Analysis. In *Modern Methods for Analysing Archaeological and Historical Glass*; Janssens, K., Ed.; Wiley: Hoboken, NJ, USA, 2013. [CrossRef]
17. Picollo, M.; Bacci, M.; Cucci, C.; Galeotti, M.; Radicati, B.; Trumpy, G. Fluorescence for the analysis of paintings. In *Optical Technologies in the Humanities*; Springer Series in Optical Sciences; Springer: Cham, Switzerland, 2019; Volume 228. [CrossRef]
18. Thoury, M.; Elias, M.; Frigerio, J.M.; Barthou, C. Non-destructive identification of varnishes by UV fluorescence spectroscopy. In *Optical Methods for Arts Archaeology*; SPIE: Bellingham, WA, USA, 2005; Volume 5857. [CrossRef]
19. Vascotto, V.; Galeotti, M.; Picollo, M.; Bacci, M. A study of UV fluorescence emission of painting materials. In Proceedings of the 10th International Conference on Non-Destructive Testing and Microanalysis for the Diagnostics and Conservation of Cultural and Environmental Heritage, Antwerp, Belgium, 8–10 June 2004.
20. Measday, D.; Walker, C.; Pemberton, B. A Summary of Ultra-Violet Fluorescent Materials Relevant to Conservation. AICCM National Newsletter, No. 137. March 2017. Available online: https://aiccm.org.au/network-news/summary-ultra-violet-fluorescent-materials-relevant-conservation/ (accessed on 10 December 2024).
21. Webb, E.K. UV-Induced Visible Luminescence for Conservation Documentation. In *UV-Vis Luminescence Imaging Techniques/Técnicas de Imagen de Luminiscencia UV-Vis*; Picollo, M., Stols-Witlox, M., Fuster-López, L., Eds.; Smithsonian Institution: Washington, DC, USA, 2019; pp. 35–60. Available online: https://hdl.handle.net/10088/105076 (accessed on 10 December 2024).
22. Hickey-Friedman, H. Ultraviolet Light Examination and Photography: Operational and Interpretive Guidelines for Art Conservators. *Objects Spec. Group Postprints* **2002**, *9*, 161–170. Available online: https://resources.culturalheritage.org/osg-postprints/v09/hickey-friedman2/ (accessed on 10 December 2024).
23. Janssens, K.; Van der Snickt, G.; Vanmeert, F.; Legrand, S.; Nuyts, G.; Alfeld, M.; Monico, L.; Anaf, W.; De Nolf, W.; Vermeulen, M.; et al. Non-Invasive and Non-Destructive Examination of Artistic Pigments, Paints, and Paintings by Means of X-Ray Methods. In *Analytical Chemistry for Cultural Heritage. Topics in Current Chemistry Collections*; Mazzeo, R., Ed.; Springer: Cham, Switzerland, 2016. [CrossRef]
24. Brocchieri, J.; Scialla, E.; D'Onofrio, A.; Sabbarese, C. Combining XRF, Multispectral Imaging and SEM/EDS to Characterize a Contemporary Painting. *Quantum Beam Sci.* **2023**, *7*, 13. [CrossRef]
25. Monico, L.; Vanmeert, F.; Janssens, K.; Abakumov, A.; Lu, Y.; Tian, H.; Liu, Y.; Yang, Y.; Chen, L.; Sanyova, J. The use of lead-based pigments and their degradation pathways in historical paintings. *Corros. Sci.* **2018**, *146*, 53–65. [CrossRef]
26. La Russa, M.F.; Ruffolo, S.A.; Belfiore, C.M.; Comite, V.; Casoli, A.; Berzioli, M.; Nava, G. A scientific approach to characterizing the painting technique of an artist: The case of Raffaele Rinaldi. *Appl. Phys. A* **2014**, *114*, 733–740. [CrossRef]
27. Barnett, J.R.; Miller, S.; Pearce, E. Colour and art: A brief history of pigments. *Opt. Laser Technol.* **2006**, *38*, 445–453. [CrossRef]
28. Amat, A.; Rosi, F.; Miliani, C.; Sassi, P.; Paolantoni, M.; Fantacci, S. A combined theoretical and experimental investigation of the electronic and vibrational properties of red lead pigment. *J. Cult. Herit.* **2020**, *46*, 374–381. [CrossRef]
29. Hradil, D.; Hradilová, J.; Lanterna, G.; Galeotti, M.; Holcová, K.; Jaques, V.; Bezdička, P. Clay and alunite-rich materials in the painting grounds of major Italian masters: Caravaggio and Mattia Preti. *Appl. Clay Sci.* **2020**, *185*, 105412. [CrossRef]
30. Zumbühl, S.; Zindel, C. Historical Oil Painting Siccatives and Applications in Oil Painting. Painting Best Practices, 2021. Available online: https://paintingbestpractices.com (accessed on 10 December 2024).
31. Hradil, D.; Grygar, T.; Hradilová, J.; Bezdicka, P. Clay and iron oxide pigments in the history of painting. *Appl. Clay Sci.* **2003**, *22*, 223–236. [CrossRef]
32. Barata, C.; Cruz, A.J.; Carballo, J.; Araújo, M.E. The Materials and Techniques of Polychrome Baroque Wooden Sculpture: Three Works From Baião, Portugal. *Int. J. Conserv. Sci.* **2010**, *1*, 19–26.
33. Miguel, C.; Pinto, J.; Clarke, M.; Melo, M.J. The alchemy of red mercury sulphide: The production of vermilion for medieval art. *Dye. Pigment.* **2014**, *102*, 210–217. [CrossRef]
34. Radepont, M.; Coquinot, Y.; Janssens, K.; Ezrati, J.-J.; de Nolf, W.; Cotte, M. Thermodynamic and experimental study of the degradation of the red pigment mercury sulfide. *J. Anal. At. Spectrom.* **2015**, *30*, 599–612. [CrossRef]
35. Keune, K.; Boon, J.J. Analytical imaging studies clarifying the process of the darkening of vermilion in paintings. *Anal. Chem.* **2005**, *77*, 4742–4750. [CrossRef] [PubMed]
36. Anaf, W.; Janssens, K.; de Nolf, W.; Cotte, M. Formation of metallic mercury during photodegradation/photodarkening of α-HgS: Electrochemical evidence. *Angew. Chem. Int. Ed.* **2013**, *52*, 12568–12571. [CrossRef] [PubMed]

37. Radepont, M. Understanding of Chemical Reactions Involved in Pigment Discoloration, in Particular in Mercury Sulfide (HgS) Blackening. Ph.D. Thesis, Université Pierre et Marie Curie (Paris VI), Paris, France, 2013. Available online: https://www.researchgate.net/publication/281659746 (accessed on 10 December 2024).
38. Cardell, C.; Guerra, I. Direct evidence for metallic mercury causing photo-induced darkening of red cinnabar tempera paints. *Commun. Chem.* **2021**, *4*, 174. [CrossRef]
39. *UNI 11118:2004*; Criteri per l'Identificazione delle Specie Legnose. UNI: Milan, Italy, 2004.
40. Schweingruber, F.H. *Mikroskopische Holzanatomie*; EAFV, Kommissionsverlag: Teufen, Switzerland, 1982; pp. 1–226.
41. Schweingruber, F.H. *Anatomie Europaeischer Hoelzer*; WSL FNP: Stuttgart, Germany, 1990; pp. 1–800.
42. Castelletti, L.; De Carlo, M. Studio silotomico delle sculture di Ettore Ferrata, in Casati L (a cura di) Omaggio ai maestri intelvesi. Ercole Ferrata, Carlo Innocenzo Carloni, Sculture e dipinti dal Museo Dioceano di Scaria Intelvi. *Quad. Pinacoteca Civ. Como* **2010**, *5*, 82–96.
43. Signorini, G.; Di Giulio, G.; Fioravanti, M. *Il Legno nei Beni Culturali, Guida alla Determinazione delle Specie Legnose*; Aguaplano: Passignano sul Trasimeno, Italy, 2014; pp. 1–344.
44. Gasanova, S.; Bakirtzis, N.; Levif-Martos, D.; Hermon, S. Giovanni Baronzio's 'Crucifixion': Analytical approaches and art historical considerations. *Herit. Sci.* **2020**, *8*, 99. [CrossRef]

heritage

MDPI

Article

Cross-Dating in Archaeology: A Comparative Archaeomagnetic, Thermoluminescence and Radiocarbon Dating of an Ancient Kiln, Ceva, Northern Italy

Evdokia Tema [1,2,*], Georgios S. Polymeris [3], Marco Casola [4] and Simone Giovanni Lerma [5]

1 Dipartimento di Scienze della Terra, Università degli Studi di Torino, Via Valperga 35, 10125 Torino, Italy
2 CIMaN-ALP Alpine Laboratory of Palaeomagnetism, Via G.U. Luigi Massa 6, 12016 Peveragno, Italy
3 Institute of Nanoscience and Nanotechnology, NCSR "Demokritos", Ag. Paraskevi, 15310 Athens, Greece; g.polymeris@inn.demokritos.gr
4 F.T. Studio S.r.l., Piazza San Domenico 2, 12016 Peveragno, Italy; marco.casola@ftstudio.eu
5 Soprintendenza Archeologia Belle Arti e Paesaggio per le Province di Alessandria, Asti e Cuneo, Cittadella, Via Pavia, 15121 Alessandria, Italy; simonegiovanni.lerma@cultura.gov.it
* Correspondence: evdokia.tema@unito.it; Tel.: +39-0116708395

Abstract

In this study, we present the dating results of an ancient kiln excavated near Ceva (Northern Italy) obtained through combined archaeomagnetic and thermoluminescence approaches. For archaeomagnetic dating, the full geomagnetic field vector (both direction and intensity) was determined. The archaeomagnetic direction was defined through stepwise alternating field demagnetization of in situ-oriented samples of baked clay, and the archaeointensity value was obtained through the Thellier–Coe protocol, including corrections for magnetic anisotropy and cooling rate effects. Thermoluminescence analyses were obtained individually on three samples, using the conventional multiple-aliquot, additive dose procedure. Archaeomagnetic dating was carried out twice, once using the directional results only and once using the full geomagnetic field vector. The independent dating provided by the thermoluminescence analysis was used for comparison, examining the added value of incorporating archaeointensity measurements alongside directional data. The new archaeomagnetic and thermoluminescence results were integrated with previously available radiocarbon dating, using Bayesian modeling for chronological reconstructions. Our results show that the use of archaeointensity in archaeomagnetic dating can be advantageous, better refining the dating. This multidisciplinary strategy underscores the significance of cross-dating in establishing robust chronological frameworks and highlights the crucial role of transdisciplinary methodologies in advancing and refining dating techniques in archaeology.

Keywords: archaeomagnetic dating; thermoluminescence; ancient kiln; combining chronologies; cross-dating

1. Introduction

Dating plays a crucial role in archaeology, particularly when historical records or age-diagnostic artifacts are absent. In such cases, laboratory-based scientific dating techniques become essential. Archaeological research has long benefited from a wide array of these methods, especially from the mid-20th century onwards, following World War II, when rapid advancements in physics and technology significantly improved the experimental protocols and the precision of such dating methods.

Radiocarbon and thermoluminescence (TL) are among the most widely used and well-established dating techniques in archaeology. These methods are particularly effective for dating ancient combustion structures, where charcoal is often preserved, and where high-temperature processes reset the TL signal to zero. Archaeomagnetic dating, though less widely known and relatively recent, has also shown strong potential in dating ancient kilns and has been successfully applied in the past [1,2]. While radiocarbon dating estimates the age of charcoal, typically resulting from incomplete wood combustion, TL and archaeomagnetic methods date the last firing of a structure, such as a kiln, effectively dating the time of its final use [3,4]. Each of these techniques relies on the ability to precisely measure a physical quantity that changes over time, and each one depends on a specific event that sets a "time-clock" to zero.

Thermoluminescence is one of the most widely recognized and effective techniques for dating heated or burnt materials [5–7]. TL dating determines the time elapsed since crystalline minerals in a material were last heated to high temperatures, as occurs in ceramics and combustion structures, or last exposed to sunlight, as in sediments or sun-dried clays. Two physical quantities are essential for TL dating: the total accumulated radiation dose over time, known as the paleodose or equivalent dose (ED, measured in Gy), and the annual dose rate (DR, measured in Gy/ka or mGy/a), which reflects the rate of energy accumulation. The age of the sample is calculated by dividing the paleodose by the annual dose rate.

Archaeomagnetic dating, on the other hand, is based on the property of certain minerals to acquire remanent magnetization when heated, recording the Earth's magnetic field at the time of their last firing [4]. These magnetic signatures, preserved in rocks, clays, and soils, and often in archaeological ceramics and baked clays, can be analyzed to reconstruct the geomagnetic field's direction and/or intensity at the time of heating. Archaeomagnetic dating is obtained by comparing this magnetic information with reference to secular variation (SV) curves that document changes in the Earth's magnetic field over time. This technique has been considerably refined in recent decades, with the development of high-resolution SV curves for several countries, and the establishment of geomagnetic field models at both regional and global scales. Most archaeomagnetic studies focus on in situ fired features, such as kilns, ovens, hearths, furnaces, and burnt floors, typically using the geomagnetic direction. However, applications based on geomagnetic intensity or the full field vector remain relatively rare [8–10], mainly due to the complexity and low success rate of archaeointensity experiments and the limited availability of reliable intensity SV curves.

In recent years, archaeomagnetic dating has become a valuable tool not only when traditional dating methods are unavailable or unsuitable but also because it allows for the direct dating of the structure itself, rather than associated materials. Notably, both TL and archaeomagnetic dating aim to date the same event: the last high-temperature firing of the artifact or structure. In contrast, radiocarbon dating targets the age of the wood from which the charcoal was formed, i.e., the time of tree cutting, which may not coincide exactly with the last use of an ancient combustion structure. Each technique, however, has its specific advantages and limitations, influenced by factors such as material availability, sample characteristics, preservation conditions, and the chronological range. When feasible, comparing and combining the results obtained from independent dating methods yields undoubtedly a more precise and robust chronological framework [11–14].

In this study, we present a cross-dating analysis of an ancient kiln excavated in Ceva, Northern Italy, by integrating new archaeomagnetic and thermoluminescence dating results with one previously obtained radiocarbon date and archaeological evidence. The inclusion of independent TL dating allowed us to evaluate the added value of incorporating archaeointensity data into archaeomagnetic analyses, alongside directional information. To our knowledge, this is the first study to successfully combine three entirely independent laboratory dating techniques applied to material from the same combustion structure. Our findings emphasize not only the benefits of cross-dating for establishing reliable chronological results but also the vital role of interdisciplinary approaches in enhancing dating methodologies.

2. Materials and Methods

2.1. Archaeological Site and Sampling

During the construction of a new rest area along the A6 Torino–Savona highway in Ceva (CN), in the locality of Mollere (44.37° N, 8.05° E), the remains of an ancient production site were identified approximately 0.6 m below ground level. The archaeological excavation, conducted by F.T. Studio s.r.l. under the direction of the Soprintendenza ABAP-AL (Ministry of Culture), revealed a complex comprising a rectangular kiln and a small service room with a square plan and a pebble foundation [15]. The kiln remains allowed for the identification of a combustion chamber (measuring 12.4 m in length and 4.3 m in width) and a *praefurnium* (measuring 4 m in length and 1–1.9 m in width), both carved into the natural clay layer of the hillside. The combustion chamber was divided longitudinally into two corridors by a central septum. A set of five pillars, which originally supported as many arches, is still visible (Figure 1a). The perforated baking floor, most likely resting on these arches, was not preserved, although some collapsed fragments were found within the kiln. The *praefurnium*, located north of the combustion chamber, had a funnel shape and a lowered canopy cover. Abundant charcoal remains, likely from the last firing, were discovered on the highly heat-altered, bluish-gray kiln floor. The kiln was excavated and shaped directly in the unfired clay, as evidenced by visible finger impressions left by its builders along the walls. Structural solidity was achieved through heat-induced hardening. The characteristics of the structure correspond to a typology used in the Italic region during the Roman era, dating back to the Republican period (V–I century BC) [16], type II c. Unfortunately, only seven non-diagnostic sherds of coarse pottery were found within the destruction layers: the presence of traces of pitch inside some of them may be linked to a ceramic typology attested in the area between the Late Iron Age ("Ligurian III C", 250–125 BC) [17] and the beginning of Roman penetration in this part of Piedmont (II–I century BC) [18]. In the same layer of destruction that filled the kiln combustion chamber, a bronze coin was also found. Although illegible, the weight and dimensions of the coin are probably attributable to a semi-uncial republican *triens*, minted between 91 and 82 BC or, alternatively, to non-Roman issues that can be dated between 130 and 90 BC [15]. Therefore, based on the general production context, the kiln can be archaeologically attributed to the late Republican period: the site bears witness to the presence of a Roman settlement along a communication route that connected the inland areas of Piedmont with the Ligurian coast at Vada Sabatia (modern-day Vado-Savona), a likely southern branch of the Via Fulvia, constructed around 125 BC [19].

For this study, a total of 22 samples of baked clay were collected: 17 from the main combustion chamber and 5 from the *praefurnium*. To be able to use the samples for archaeomagnetic directional analysis, all samples were oriented in situ with both a magnetic and a solar compass and an inclinometer. Sampling was performed either with the plaster of Paris, used to create a flat orientation surface, or with gluing small plastic disks on the baked clay (Figure 1b,c). Three of these samples were used for TL analysis. One charcoal sample was also extracted from the *praefurnium* (US29) and used for radiocarbon dating at the CEDAD Laboratory in Lecce, Italy.

Figure 1. (**a**) View of the investigated kiln at Ceva, località Mollere, with the position of the investigated samples (white dots: archaeomagnetic samples; red stars: TL samples; orange diamond: radiocarbon sample); (**b**) oriented samples collected for archaeomagnetic research with the use of plaster of Paris; (**c**) oriented samples with plastic disc.

2.2. *Archaeomagnetic Analysis*

Archaeomagnetic analyses were conducted at the ALP-CIMaN Palaeomagnetic Laboratory in Peveragno, Italy. The magnetic mineralogy of representative samples was examined using isothermal remanent magnetization (IRM) acquisition curves, obtained with an ASC pulse magnetizer, which applied IRM at increasing magnetic fields up to 1 T. A JR6 spinner magnetometer (AGICO) was used to measure the remanent magnetization after each IRM step.

The archaeomagnetic direction was determined through stepwise alternating field demagnetization, performed up to a peak field of 100–120 mT using a D2000 demagnetizer (ASC Scientific, Narragansett, RI, USA). The demagnetization results were plotted on orthogonal projection diagrams [20], and the direction of the Characteristic Remanent Magnetization (ChRM) for each sample was calculated using principal component analysis with Remasoft 3.0 software [21].

Archaeointensity was assessed using the classical Thellier method, as modified by Coe [22,23]. Samples were subjected to multiple zero-field and in-field heating and cooling steps in aTD-48 furnace (ASC Scientific, USA), up to a maximum temperature of 500–530 °C. During the in-field steps, a laboratory field of 60 μT was applied during both heating and cooling. To monitor potential mineralogical alterations, additional partial thermoremanent magnetization (pTRM) checks were performed every two experimental temperature steps. The effect of magnetic anisotropy on the thermoremanent magnetization (ATRM) was also investigated through the acquisition of six additional TRM in six different directions (+x, −x, +y, −y, +z, −z) in order to define the ATRM tensor. Finally, the cooling rate dependence of TRM acquisition was estimated by applying three additional in-field heating steps: one with rapid cooling time of around 45 min, one with slow cooling time of around 6 h, and one more with rapid cooling time of around 45 min. From these three steps, we obtained the correction factor for the effect of cooling rate upon TRM intensity, and the alteration factor used to estimate changes in the TRM acquisition capacity of the specimens during the cooling rate protocol. Archaeointensity results were plotted in Arai diagrams and analyzed using the Thellier Tool v.4.22 software [24].

2.3. Thermoluminescence Analysis

TL analyses were carried out at the Nuclear Physics Laboratory of the Physics Department, Aristotle University of Thessaloniki, Greece. ED was calculated using the multigrain, multiple-aliquot, additive dose thermoluminescence procedure (MAAD TL), as described in [5,7,25]. Handling and pretreatment of the samples were performed in subdued red-light conditions. Grains of dimensions within the range between 4 and 12 μm were suspended in acetone and finally precipitated onto 1 cm diameter aluminum discs according to [26]. The chemical protocol suggested by [27] was applied, following mechanical treatment. A 5% water content was used for the calculation of the dose rate.

All TL measurements were performed by means of a Littlemore-type 711 setup, with a PM tube (EMI 9635QA bialkali; Sb K-Cs), a thermocouple-type 90/10 Ni/Cr and 97/03 Ni/Al, and a heat filter transmitting in the 320–440 nm range. In all cases, for beta doses, a $^{90}Sr/^{90}Y$ beta source delivering 1.72 Gy/min at the sample position was used. All TL measurements were performed in a nitrogen atmosphere up to a maximum temperature of 500 °C at a constant heating rate of 2 °C/s. Three different additive doses were used, namely, 7, 14, and 20 Gy. Only linear fittings were used in the present approach. Equivalent doses were calculated with 1σ error values. Errors were mainly derived from the uncertainties in curve fitting and were calculated by standard error propagation analysis.

The annual dose rate was calculated based on the decay rates of naturally occurring radionuclides inside the clay matrix, namely, potassium (^{40}K), thorium (^{232}Th), and natural uranium (^{235}U and ^{238}U), along with cosmic rays. Uranium and thorium concentrations (in units of parts per million, ppm) were estimated using thick source alpha counting [28,29], while the potassium concentration (in %) by using Scanning Electron Microscopy coupled with Energy Dispersion X-ray (SEM-EDX) analysis [30]. Dose-rate calculations were made using the conversion factors of [31]. Only the clay matrix was measured for the dose rate calculations.

3. Results

3.1. Direction and Archaeointensity Determination

Magnetic mineralogy investigated through stepwise IRM curves shows the presence of low coercivity minerals, most probably magnetite and/or Ti-magnetite, together with a contribution of high-coercivity minerals, most probably hematite. These results are further confirmed by the intensity decay curves obtained through stepwise AF demagnetization, which show that about 70 to 80% of the Natural Remanent Magnetization (NRM) is canceled at applied fields of 100 mT, while in some cases (e.g., sample 19) an important portion of the NRM is not demagnetized (Figure 2a,b). In general, almost all samples show a very stable and easily isolated ChRM, with linear and well-defined Zijdelveld diagrams (Figure 2c). The mean ChRM direction, calculated from 18 independently oriented samples, is D = 359.1° I = 64.4 °, alpha-95 = 2.5°, and precision parameter k = 192 (Figure 2d).

Archaeointensity results show generally well-defined and linear Arai plots, with successful pTRM checks (Figure 3). From the 17 specimens investigated, coming from 6 independent samples (generally three specimens per sample were investigated), 15 gave successful archaeointensity determinations. Archaeointensity values at the specimen level were accepted only when the linear segment in the Arai plots was defined by more than four points ($N > 6$), quality parameter q was higher than five (generally $q > 15$), and no clear magnetomineralogical alteration was observed. All results were corrected for anisotropy and cooling-rate effects. The application of these corrections resulted in a reduction in the initial archaeointensity value by approximately 10%, underscoring their significance in ensuring the reliability and accuracy of archaeointensity determinations. The final mean archaeointensity value for the kiln calculated after the cooling rate and anisotropy corrections, based on the mean values of the six independently collected samples, is F = 61.6 ± 3.7 μT. The detailed archaeointensity results, at both specimen and sample levels, are given in the Supplementary Materials (Table S1).

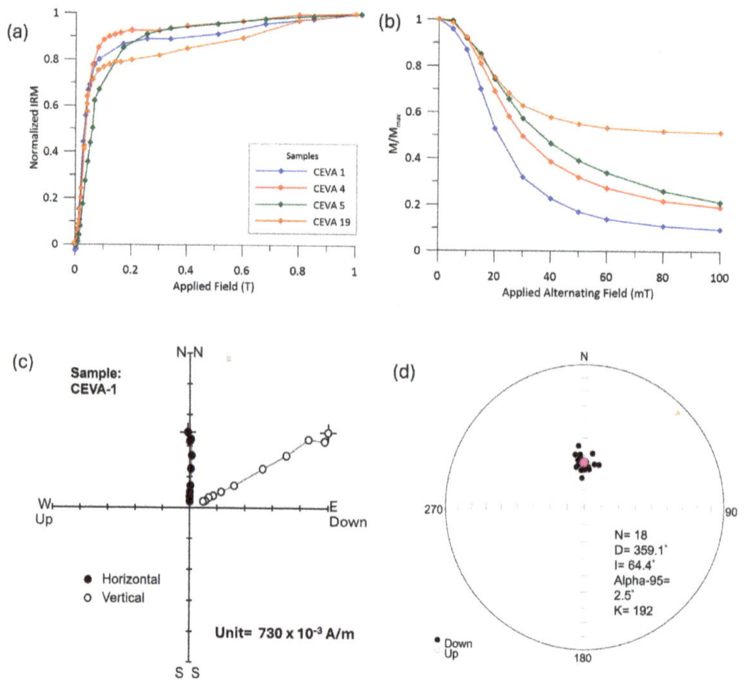

Figure 2. Archaeomagnetic results: (**a**) IRM acquisition curves and (**b**) intensity decay curves for representative samples; (**c**) stepwise AF demagnetization results represented in Zijderveld diagram; (**d**) equal area projection of the ChRM direction of the investigated samples. The pink dot illustrates the mean direction calculated for the kiln.

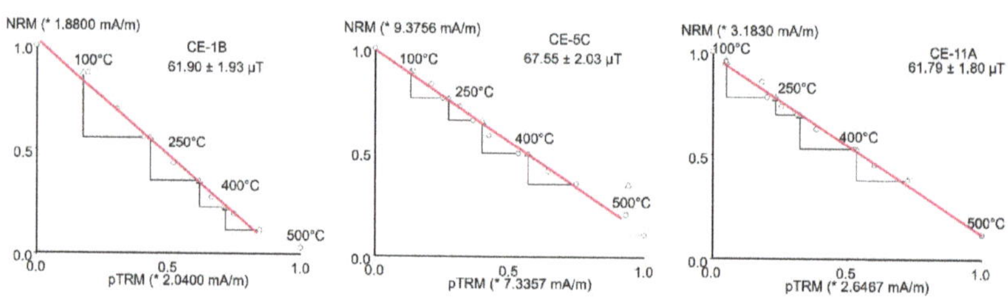

Figure 3. Archaeointensity results obtained for representative samples and represented in NRM-TRM diagrams, known as Arai plots. Diagrams are normalized to the initial NRM intensity. During the archaeointensity experiments, regular pTRM checks were performed (triangles in the Arai plots).

3.2. Archaeomagnetic Dating

Archaeomagnetic dating of the last firing of the Ceva kiln was obtained using the ArchaeoPyDating tool [32], which is an online, open-access web-based interface that allows the comparison of the experimentally determined directional and intensity values from the investigated material with a reference SV curve available for a certain geographical area. In this study, we calculated a temporal probability density separately for each geomagnetic field element (declination, inclination, and intensity), after comparison with the most recently published Italian secular variation curve [33], calculated at the locality of Viterbo (Figure 4a). Archaeomagnetic dating was conducted twice: first using only directional data, and then using the full geomagnetic vector (combining both direction and intensity). Comparison of the directional data with the Italian SV curves [33] yielded two possible age intervals at a 95% probability level: 435–370 BC and 260 BC–35 AD (Figure 4b–d). Based on the stratigraphy and the available archaeological evidence, which suggests a late Republican period for the findings, the earlier interval can be confidently excluded, suggesting that the final firing of the Ceva kiln most likely occurred between 260 BC and 35 AD. Dating was then repeated for a second time, this time incorporating the full geomagnetic vector (direction and intensity). This yielded a better constrained chronological range, placing the last use of the kiln in a very narrow and well-defined time interval, between 115 BC and 45 AD (35 BC ± 80) (Figure 4e–h).

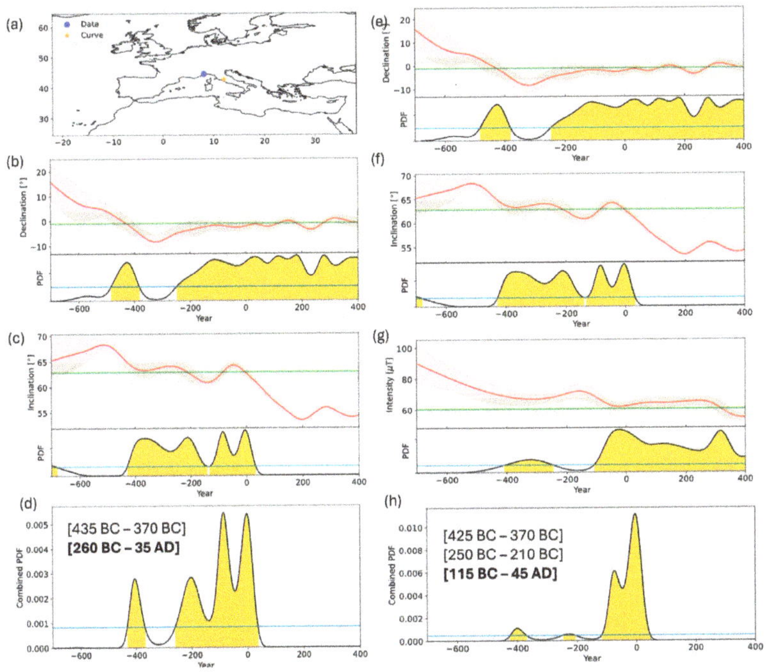

Figure 4. Archaeomagnetic dating results obtained after comparison with the Italian SV curves (Tema and Lanos, 2021) [33]: (**a**) Location of the Ceva kiln (blue dot) and origin coordinates of the Italian reference curve (Viterbo, yellow star); (**b**,**c**) Comparison of the declination and inclination values (in green) with the declination and inclination reference SV curves (in red), respectively. Bands represent the uncertainties of the laboratory measurements and the curves; (**d**) combined declination and inclination probability densities (from (**b**,**c**)), representing the final dating result based on the directional data; (**e**,**f**) same as (**b**,**c**); (**g**) comparison of the intensity value with the intensity reference curve; (**h**) combined directional and intensity probability densities (from (**e**–**g**)), representing the final dating result based on the full geomagnetic field vector (direction and intensity). See text for further details.

3.3. *Thermoluminescence Dating*

TL results for the three independent samples are reported in Table 1. TL glow curves and a typical plot of ED against glow curve temperature for representative samples are shown in Figure 5. In all cases, ED plateaus are wide enough, over 105 °C wide (Figure 5a and Table 1). Linearity was established using the slope curve S over the entire TL glow curve temperature region [25], as clearly seen in both insets of Figure 5b. The equivalent doses were obtained as the mean values of the best plateaus for each sample. The calculated ages for the three samples investigated are very consistent among each other, providing an average age of 137 BC to 181 AD, at 95% probability. Insignificant fading (<1%) was detected over a period of one month.

Table 1. Summary of the TL dating results for the three samples investigated. Error values are within 1σ level for all values except the TL ages, which are presented within 2σ error level.

Sample	U (ppm)	^{232}Th (ppm)	^{40}K (%)	ED (Gy)	ΔT (°C)	DR (Gy/ka)	TL Age (ka)	TL Age (yrs BC)
CE2	7.01 ± 0.10	11.14 ± 0.19	0.89 ± 0.05	7.93 ± 0.33	125 ± 2	3.736 ± 0.257	2.120 ± 0.203	98 ± 203
CE5	7.44 ± 0.11	12.18 ± 0.15	0.95 ± 0.06	8.36 ± 0.32	108 ± 2	3.989 ± 0.231	2.095 ± 0.198	71 ± 198
CE20	7.22 ± 0.11	11.44 ± 0.13	0.97 ± 0.05	8.04 ± 0.29	114 ± 2	3.893 ± 0.212	2.066 ± 0.206	41 ± 206

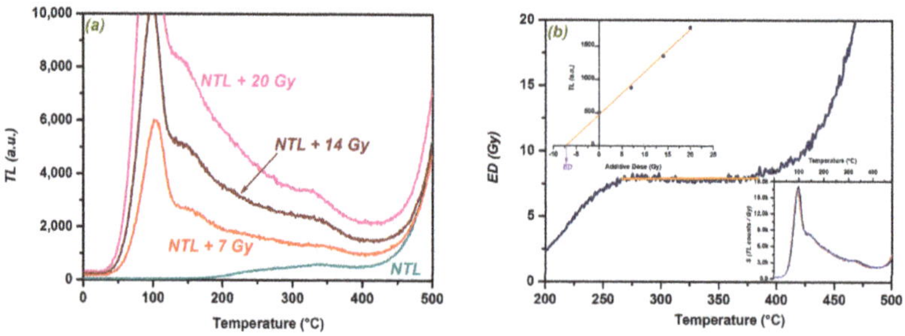

Figure 5. Thermoluminescence results: (**a**) TL glow curves related to the TL age calculation for the sample CE5. The additive doses delivered were 7, 15, and 20 Gy. Reheats have been subtracted. Each glow curve plotted is the average of three individually measured glow curves. (**b**) The equivalent dose plotted vs. temperature for the sample CE2. Solid line indicates the best ED plateau. Upper inset presents a representative NTL + β (filled points) dose response curve for the temperature of 300 °C. The arrow shows the equivalent dose. Lower inset presents the linearity curve S throughout the entire TL glow curve temperature regions for all three additive doses.

4. Discussion and Conclusions

One of the unique strengths of comparing TL and archaeomagnetic dating results lies in their ability to independently date the same physical event, which corresponds to the last high-temperature firing of a structure or object. This overlap is especially valuable in fired archaeological structures, such as the kiln investigated in our study, where materials have been repeatedly heated at high temperatures during the continuous kiln's use. In such cases, comparing results from these two methods offers several key advantages for refining and validating chronological interpretations, while at the same time, multidisciplinary dating cross-checking strongly contributes to overcoming the specific technical and methodological limitations related to each single dating technique.

Like all dating techniques, TL has specific limitations mainly related to the sensitivity to environmental radiation and to the relatively high uncertainty of the final result. Accurate dose rate estimation requires detailed knowledge of the surrounding soil's radioactivity over time, which may not always be possible, while clay matrix inhomogeneity induced by repeated heatings can result in age overestimations [14]. Another possible error source could be the leaching of the potassium content [34]. On the other hand, archaeomagnetic dating is importantly limited by the availability and resolution of well-constrained reference SV curves, which are more developed in some parts of the world than others, and better detailed for some chronological periods than others [35]. Moreover, determining the direction of the past geomagnetic field requires that the material is still found in situ (at the exact position it was heated for the last time), while laboratory archaeointensity determination procedures are complicated and time-consuming, with often a low success rate. Post-depositional movements or re-heating of the material can alter the magnetic signal, complicating interpretations.

In this perspective, and given the individual limitations of each method, integrating multiple dating approaches as proposed in this study, undoubtedly enhances chronological precision and confidence, while at the same time, such cross-checking offers the opportunity to test and improve the single techniques. In the case of the Ceva kiln, apart from the new TL and archaeomagnetic data presented here, radiocarbon analysis was previously carried out on one charcoal sample (LTL21346) collected from the stratigraphic level US29 from the *praefurnium* of the kiln. Since this charcoal-rich layer lies immediately beneath the remains of the destruction of the *praefurnium* covering, it is likely that it consists of the combustion remains of the last firing. Sample preparation and measurements were performed at the CEDAD Laboratory in Lecce (Centro di Datazione e Diagnostica), and the radiocarbon dating was obtained using the accelerator mass spectroscopy (AMS) technique. The result was corrected for δ13C [36] and calibrated to calendar age [37], using the OxCal3.10 software (the uncalibrated age is 2134 ± 45 BP). According to the report, the calibrated radiocarbon age of the sample calculated at 95% of probability is 355 BC–279 BC (19.9%) and 232 BC–42 BC (75.5%) (CEDAD Technical report, 2021: Rif. CEDAD: 2021_00116, see Supplementary Materials Table S2). Thanks to this radiocarbon result, the Ceva kiln represents a unique case study so far in the literature, giving the possibility to directly cross-check three of the most widely used dating techniques in archaeology: TL, archaeomagnetism, and radiocarbon.

Our new TL and archaeomagnetic results show exceptionally good agreement. Based on the TL dating, the last heating of the kiln occurred within the time range between 137 BC to 181 AD (22 ± 159 AD) within a 95.4% confidence level (2σ). On the other hand, the archaeomagnetic dating based on the directional results gives an age of 260 BC and 35 AD. It is interesting to note that using the full geomagnetic field vector for archaeomagnetic dating purposes (instead of direction only) further constrains the archaeomagnetic age at 115 BC to 45 AD, being in excellent agreement with the TL results. These results are also in agreement with the radiocarbon date, even if radiocarbon suggests a slightly older age, most probably due to the well-known old wood effect, taking into consideration that radiocarbon dates the age of the investigated wood and not the time of the kiln's abandonment.

Having this exceptional opportunity to have dating results involving three different dating methods, we applied Bayesian modelling to integrate TL, archaeomagnetic, and radiocarbon dates into a unique age. Analysis was performed using the ChronoModel software v. 3.2.7 [38,39], which provides a robust Bayesian tool for chronology building in order to estimate the date of a target event by a combination of individual dates of various techniques obtained from archaeological materials, assuming the samples are contemporaneous. ChronoModel is open-source and freely available [40], ensuring full transparency, repeatability, and reproducibility of the analyses. Its calculations are based on Markov Chain Monte Carlo (MCMC) numerical methods, providing a statistically sound framework for estimating the timing of archaeological events.

The comparison of all dating results confirms their very good agreement, while the combined Bayesian analysis constrains the last use of the kiln to between 213 BC and 111 AD at a 2σ confidence level (Figure 6). This interval is fully consistent with the archaeological evidence and, although broader than those obtained from the individual techniques, it provides a more reliable chronological framework and emphasizes the importance of cross-checking dating methods in archaeology. Extremely narrow dating intervals are not necessarily more reliable, particularly when derived from a single method, which may give a misleading impression of precision.

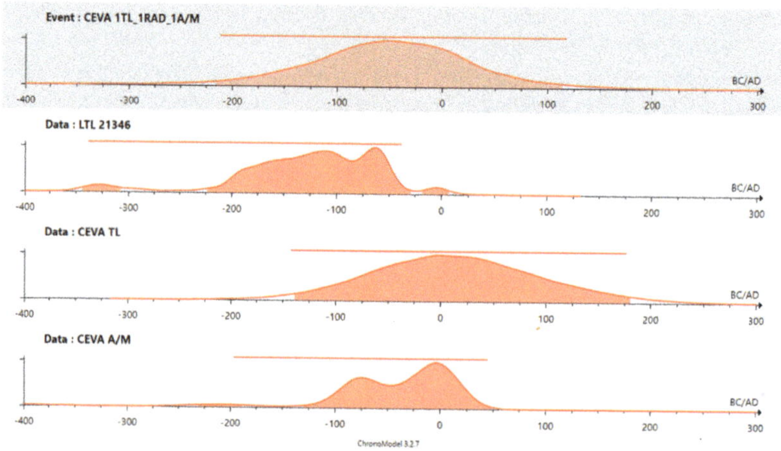

Figure 6. Bayesian analysis combining the ages obtained by all applied techniques, namely, TL, archaeomagnetism (AM), and radiocarbon (sample LTL 21346), using Chronomodel software. Red lines correspond to the individual probability density calculated for each technique, while the red areas present the most significant density in the framework of the combined analysis procedure. The final combined interval at 95% of probability is reported at the upper part of the figure (indicated with grey shadow).

Our study demonstrates that, although laboratory-based dating methods such as TL, archaeomagnetism, and radiocarbon dating each have inherent limitations, their strategic integration and cross-validation are essential for building robust archaeological chronologies. Moreover, the inclusion of archaeointensity measurements in archaeomagnetic dating proves particularly beneficial, as it helps to refine age estimates and achieve tightly constrained results. Undoubtedly, integrative approaches not only compensate for the limitations of individual techniques but also contribute to a more detailed and reliable understanding of the archaeological record.

Supplementary Materials: The following supporting information can be downloaded at: https://www.mdpi.com/article/10.3390/heritage8090358/s1, Table S1: Detailed archaeointensity results; Table S2: Radiocarbon results obtained for the charcoal sample US29-C1.

Author Contributions: E.T.: Conceptualization; Methodology; Sampling; Laboratory analysis; Data interpretation; Data curation; Writing—Original Draft. G.S.P.: Methodology; Laboratory analysis; Data interpretation; Data curation; Writing—Review & Editing. M.C.: Sampling; Resources; Data curation; Visualization; Writing—Review & Editing. S.G.L.: Resources; Visualization; Writing—Review & Editing. All authors have read and agreed to the published version of the manuscript.

Funding: This study was partially funded by F.T. Studio s.r.l. and Autostrada dei Fiori S.p.A.

Data Availability Statement: All data are freely available upon request.

Acknowledgments: We warmly acknowledge AIAr—Associazione Italiana di Archeometria for fully covering the costs of this Open Access publication. We also thank F.T. Studio s.r.l. and Autostrada dei Fiori S.p.A. for their financial support in funding the excavation and part of the analysis. Enzo Ferrara is also warmly acknowledged for his valuable assistance during the sampling process.

Conflicts of Interest: The authors declare no conflicts of interest.

References

1. Tema, E.; Fantino, F.; Ferrara, E.; Lo Giudice, A.; Morales, J.; Goguitchaichvili, A.; Camps, P.; Barello, F.; Gulmini, M. Combined archaeomagnetic and thermoluminescence study of a brick kiln excavated at Fontanetto Po (Vercelli northern Italy). *J. Archaeol. Sci.* **2013**, *40*, 2025–2035. [CrossRef]
2. Tema, E.; Fantino, F.; Ferrara, E.; Allegretti, S.; Lo Giudice, A.; Re, A.; Barello, F.; Vella, S.; Cirillo, L.; Gulmini, M. Archaeological archaeomagnetic and thermoluminescence investigation of a baked clay kiln excavated at Chieri, northern Italy: Contribution to the rescue of our cultural heritage. *Ann. Geophys.* **2014**, *57*, G0548. [CrossRef]
3. Becker, H.; Göksu, H.Y.; Regulla, D.F. Combination of archaeomagnetism and thermoluminescence for precision dating. *Quaternary Sci. Rev.* **1994**, *13*, 563–567. [CrossRef]
4. Batt, C. Archaeomagnetic Dating. In *Handbook of Archaeological Sciences*; Pollard, A.M., Armitage, R.A., Makarewicz, C.A., Eds.; Wiley: Hoboken, NJ, USA, 2023. [CrossRef]
5. Aitken, M.J. *Thermoluminescence Dating*; Academic Press: London, UK, 1985.
6. Aitken, M.J. *Science-Based Dating in Archaeology*; Longman Press House: London, UK; New York, NY, USA, 1990.
7. Wagner, G.A. *Age Determination of Young Rocks and Artifacts: Physical and Chemical Clocks in Quaternary Geology and Archaeology*; Springer: Berlin/Heidelberg, Germany, 1998.
8. Carrancho, Á.; Goguitchaichvili, A.; Morales, J.; Espinosa-Soto, J.A.; Villalaín, J.J.; Arsuaga, J.L.; Baquedano, E.; Pérez-González, A. Full-Vector Archaeomagnetic Dating of a Medieval Limekiln at Pinilla Del Valle Site (Madrid, Spain). *Archaeometry* **2017**, *59*, 373–394. [CrossRef]
9. Principe, C.; Goguitchaichvili, A.; Devidze, M.; La Felice, S.; Cejudo, R.; Morales, J.; Cantini, F. Archaeomagnetic Dating of Three Furnaces inside the Middle Age Settlement of San Genesio (San Miniato Pisa, Italy). *Land* **2022**, *11*, 1936. [CrossRef]
10. Tema, E.; Camps, P.; Ferrara, E.; Poidras, T. Directional results and absolute archaeointensity determination by the classical Thellier and the multispecimen DSC protocols for two kilns excavated at Osterietta Italy. *Stud. Geophys. Geod.* **2015**, *59*, 554–577. [CrossRef]
11. Tema, E.; Polymeris, G.; Morales, J.; Goguitchaichvili, A.; Tsaknaki, V. Dating of ancient kilns: A combined archaeomagnetic and thermoluminescence analysis applied to a brick workshop at Kato Achaia, Greece. *J. Cult. Herit.* **2015**, *16*, 496–507. [CrossRef]
12. Tema, E.; Ferrara, E.; Angelici, D.; Fantino, F.; Panero, E. The importance of multidisciplinary dating in rescue excavations: The case of Santhià Northern Italy. *J. Archaeol. Sci. Rep.* **2019**, *28*, 102059. [CrossRef]
13. Kondopoulou, D.; Aidona, E.; Ioannidis, N.; Polymeris, G.S.; Tsolakis, S. Archaeomagnetic study and thermoluminescence dating of Protobyzantine kilns (Megali Kypsa, North Greece). *J. Archaeol. Sci. Rep.* **2015**, *2*, 156–168. [CrossRef]
14. Aidona, E.; Polymeris, G.; Camps, P.; Kondopoulou, D.; Ioannidis, N.; Raptis, K. Archaeomagnetic versus Luminescence Methods: The case of an Early Byzantine Ceramic Workshop in Thessaloniki. *Greece. Arch. Anthropol. Sci.* **2018**, *10*, 725–741. [CrossRef]
15. Lerma, S.G.; Casola, M.; Barello, F.; Tema, E. Ceva frazione Mollere, autostrada A6 Torino-Savona. Realizzazione area di sosta di Priero Sud. Fornace per laterizi di età romana. *Quad. Archeol. Del Piemonte* **2023**, *7*, 165–169.
16. Cuomo Di Caprio, N. Proposta di classificazione delle fornaci per ceramica e laterizi nell'area italiana dalla preistoria a tutta l'epoca romana. *Sibrium Cent. Studi Preist. Archeol. Varese Italy* **1971–1972**, *11*, 371–464.
17. Ferrero, L.; Giaretti, M.; Padovan, S. Gli abitati della Liguria interna: La ceramica domestica. In *Ligures Celeberrimi, la Liguria Interna Nella Seconda età del Ferro: Atti del Convegno Internazionale (Mondovì, 26–28 Aprile 2002)*; Istituto Internazionale di Studi Liguri: Bordighera, Italy, 2004; pp. 51–80.
18. Ferrero, L. Insediamenti e popolamento dell'*Ager* di *Augusta Bagiennorum* dall'età del Ferro ai primi secoli dell'Impero. In *Augusta Bagiennorum. Storia e Archeologia di una Città Augustea*; CELID: Torino, Italy, 2014; pp. 39–59.
19. Negro Ponzi Mancini M., M. Strade e insediamenti nel cuneese dall'età romana al Medioevo: Materiali per lo studio della struttura del territorio. *Boll. Soc. St. Stor. Arch. ed Art. Prov. Cuneo* **1981**, *85*, 8–84.
20. Zijderveld, J. AC demagnetization of rocks: Analysis of results. In *Methods in Paleomagnetism Collinson*; Collinson, D.W., Creer, K.M., Runcorn, S.K., Eds.; Elsevier: New York, NY, USA, 1967; pp. 254–256.
21. Chadima, M.; Hrouda, F. Remasoft 3.0 a user-friendly paleomagnetic data browser and analyzer. *Trav. Geophys.* **2006**, *XXVII*, 20–21.
22. Thellier, E.; Thellier, O. Sur l'intensité du champ magnétique terrestre dans le passé historique et géologique. *Ann. Geophys.* **1959**, *15*, 285–376.

23. Coe, R.S.; Grommé, S.; Mankinen, A. Geomagnetic paleointensities from radiocarbon dated lava flows on Hawai and the question of the Pacific nondipole low. *J. Geophys. Res.* **1978**, *83*, 1740–1756. [CrossRef]
24. Leonhardt, R.; Heunemann, C.; Krása, D. Analysing absolute paleointensity determinations: Acceptance criteria and the software Thelliertool4.0. Geochem. *Geophys. Geosyst.* **2004**, *5*. [CrossRef]
25. Polymeris, G.S. Thermoluminescence as a Tool for Age Assessment and Palaeothermometry Studies of Baked Clayey Artefacts. In *Ceramic Analysis: Laboratory Methods*; Hein, I., Ed.; Cambridge Elements, Current Archaeological Tools and Techniques; Cambridge University Press: Cambridge, UK, 2025; Volume 2025.
26. Zimmerman, D.W. Luminescence dating using fine grains from pottery. *Archaeometry* **1971**, *13*, 29–56. [CrossRef]
27. Vieillevigne, E.; Guibert, P.; Bechtel, F. Luminescence chronology of the medieval citadel of Termez, Uzbekistan: TL dating of bricks masonries. *J. Archeol. Sci.* **2007**, *34*, 1402–1416. [CrossRef]
28. Hossain, S.M.; De Corte, F.; Vandenberghe, D.; van den haute, P. A comparison of methods for the annual radiation dose determination in the luminescence dating of loess sediment. *Nucl. Inst. Methods Phys. Res. A* **2002**, *490*, 598–613. [CrossRef]
29. Liritzis, I.; Vafiadou, A. Calibration aspects of thick source alpha counter ZnS system. *Measurement* **2012**, *45*, 1966–1980. [CrossRef]
30. Liritzis, I.; Mavrikis, D.; Zacharias, N.; Sakalis, A.; Tsirliganis, N.; Polymeris, G.S. Potassium determinations using SEM, FAAS and XRF: Some experimental notes. *Mediter. Archaeol. Archaeom.* **2011**, *11*, 169–178.
31. Liritzis, I.; Stamoulis, K.; Papachristodoulou, C.; Ioannides, K. A re-evaluation of radiation dose rate conversion factors. *Mediterr. Arch. Archaeomet.* **2013**, *13*, 1–15.
32. Serrano, M.; Pavón-Carrasco, F.J.; Campuzano, S.A.; Osete, M.L. ArchaeoPyDating: A new user-friendly release for archaeomagnetic dating. *Archaeometry* **2024**, *66*, 1424–1437. [CrossRef]
33. Tema, E.; Lanos, P. New Italian directional and intensity archaeomagnetic reference curves for the past 3000 years: Insights on secular variation and implications on dating. *Archaeometry* **2021**, *63*, 428–445. [CrossRef]
34. Zacharias, N.; Buxeda i Garrigos, J.; Mommsen, H.; Schwedt, A.; Kilikoglou, V. Implications of burial alterations on luminescence dating of archaeological ceramics. *J. Archaeol. Sci.* **2005**, *32*, 49–57. [CrossRef]
35. Casas, L.; Tema, E. Investigating the expected archaeomagnetic dating precision in Europe: A temporal and spatial analysis based on the SCHA.DIF.3K geomagnetic field model. *J. Archaeol. Sci.* **2019**, *108*, 104972. [CrossRef]
36. Stuiver, M.; Polach, H.A. Discussion Reporting of 14C Data. *Radiocarbon* **1977**, *19*, 355–363. [CrossRef]
37. Reimer, P.J.; Bard, E.; Bayliss, A.; Beck, J.W.; Blackwell, P.G.; Bronkamsey, C.; Buck, C.E.; Cheng, H.; Edwards, R.L.; Friedrich, M.; et al. IntCal13 and Marine13 radiocarbon age calibration curves 0–50,000 years cal BP. *Radiocarbon* **2013**, *55*, 1869–1887. [CrossRef]
38. Lanos, P.; Philippe, A. Hierarchical Bayesian modeling for combining dates in archeological context. *J. Société Française Stat.* **2017**, *158*, 17.
39. Lanos, P.; Philippe, A. Event date model: A robust Bayesian tool for chronology building. *Commun. Stat. Appl. Methods* **2018**, *25*, 131–157. [CrossRef]
40. Lanos, P.; Dufresne, P. Chronomodel Version 2.0: Software for Chronological Modelling of Archaeological Data Using Bayesian Statistics. 2019. Available online: http://www.chronomodel.com.

heritage

MDPI

Article

Advanced Isotopic Techniques to Investigate Cultural Heritage: The Research Activities at the iCONa Laboratory

Noemi Mantile *, Simona Altieri, Maria Rosa di Cicco, Valentina Giacometti and Carmine Lubritto *

Dipartimento di Scienze e Tecnologie Ambientali Biologiche e Farmaceutiche, Mediterranean bioArchaeological Research Advances (MAReA) Centre, Università degli Studi della Campania "Luigi Vanvitelli", Via Vivaldi 43, 81100 Caserta, Italy; simona.altieri@unicampania.it (S.A.); mariarosa.dicicco@unicampania.it (M.R.d.C.); valentina.giacometti@unicampania.it (V.G.)

* Correspondence: noemi.mantile@unicampania.it (N.M.); carmine.lubritto@unicampania.it (C.L.)

Abstract

Isotopic analyses are useful tools with a wide range of applications, including environmental studies, archaeology and biomedicine. Founded in 2019 at the University of Campania "Luigi Vanvitelli", the iCONa laboratory specialises in stable isotope mass spectrometry, with a particular focus on cultural heritage. The laboratory performs carbon, nitrogen and oxygen isotopic analyses, including the most recent advances in compound-specific stable isotope analysis of amino acids (CSIA-AAs). In addition to these analytical services, iCONa provides chemical and physical sample preparation for a variety of sample types. This paper focuses on our applications in the field of cultural heritage, exploring how the analysis of stable isotopes performed on archaeological remains can be used to reconstruct past subsistence strategies and human behaviours. We also discuss the challenges inherent in isotopic analysis and recent methodological advances in the field.

Keywords: stable isotope analysis; cultural heritage; IRMS; compound-specific analysis; dietary mixing models; elemental analysis; incremental dentine

1. Introduction

Isotopic techniques have become powerful and versatile tools across a wide range of disciplines, including environmental and ecological sciences [1], archaeology [2], food traceability [3,4], biomedicine [5] and forensic investigations [6,7]. At the core of these applications lies the concept of isotopes, i.e., atoms of the same element which differ in the number of neutrons inside their nucleus [8]. In stable isotope analysis, the isotopic composition of an element is typically expressed as the ratio of its heavier to its lighter isotope [9]. The "delta notation" (δ) expresses the relative difference in this ratio in a sample compared to a defined standard, typically reported in parts per thousand (‰) [10].

A key advantage of isotopic analysis is its ability to differentiate between substances that are chemically identical but originate from different sources, thereby enabling the tracing of specific elements or molecules. The technique is also micro-destructive, requiring only milligram or sub-milligram quantities of material, thus minimising sample loss. This is especially relevant when analysing irreplaceable materials, such as archaeological specimens [11–13].

In the framework of these relevant methodologies, the iCONa lab ("isotopic Carbon, Oxygen and Nitrogen analysis laboratory") was founded in 2019 within the Department of Environmental, Biological and Pharmaceutical Sciences and Technologies at the University

https://doi.org/10.3390/heritage8080296

of Campania "Luigi Vanvitelli", with the primary function of performing stable isotope mass spectrometry analysis, with applications principally in archaeometry, environmental, food and biomedical fields. Moreover, the iCONa lab contributes to cultural heritage research within the MAReA (Mediterranean bioArchaeological Research Advances) centre, an inter-university laboratory that is jointly operated by Sapienza Università di Roma and University of Campania "Luigi Vanvitelli".

The first facilities installation took place in March 2019, with subsequent years witnessing the incorporation of additional instrumentation. This expansion aimed to enable the isotopic analysis of carbon ($\delta^{13}C$), nitrogen ($\delta^{15}N$) and oxygen ($\delta^{18}O$), as well as the determination of carbon and nitrogen concentrations (%C and %N) across various sample types, including solid, liquid and gaseous samples. A recent development has been the expansion of a specific section dedicated to compound-specific analysis, with the result that amino acids and lipids can now be measured.

In addition to the measurement of isotopic ratios, the laboratory provides a section of physical-chemical sample preparation, with specific protocols developed depending on the type of sample to be analysed. This allows organic and inorganic sample preparation also for radiocarbon (^{14}C) dating and strontium isotope analysis ($^{87}Sr/^{86}Sr$), being performed in collaboration with external research centres.

In this paper, we explore our applications in the cultural heritage field, with specific emphasis on stable isotope analysis conducted on archaeological remains, including human and faunal skeletal samples, and charred botanical materials. This technique is widely employed to investigate research questions concerning ancient subsistence strategies, environmental conditions, mobility, animal management and agricultural practices of past populations, allowing to shed light on sociocultural developments, trade networks and adaptations to environmental changes [9,14].

However, isotopic analyses present inherent challenges and limitations that can compromise the reliability of the data. Challenges associated with sampling—including representativeness, sample size, and ethical and scientific concerns—must be considered [12,15,16]. Standardising pretreatment protocols is also essential to minimise variability and contamination, supported by detailed documentation of all chemical steps to ensure methodological transparency and allowing for data reproducibility [17]. Moreover, data interpretation can vary significantly depending on the scale of analysis, which should be appropriately tailored according to the research question. Consequently, it is imperative to address these issues to ensure the validity of the analytical results and data interpretation.

Here we provide an overview of our applications in the cultural heritage field, including a description of the facilities utilised, the type of samples pretreatment and our progression towards more advanced isotopic techniques. By highlighting these elements, we aim to address the evolving methodologies that can enhance our understanding of past human behaviours through stable isotope analysis, with a specific focus on the investigation of past subsistence strategies.

2. Materials and Methods

2.1. Facilities

The laboratory is equipped with high-precision analytical instrumentation, all branded Thermo Scientific (Bremen, Germany). An Isotope Ratio Mass Spectrometer (IRMS Delta V Advantage, Thermo Fisher Scientific, Waltham, MA, USA) is used to measure isotopic ratios, based on the principle of separations of ions with different masses, permitting the measurement of their relative abundance [18]. The instrument is provided with a dual inlet system for direct measurement of gaseous samples, while is connected through a Helium flow to peripherals for solid and liquid samples, each peripheral depending on the specific isotopic ratio of interest.

For $\delta^{13}C$ and $\delta^{15}N$ analysis, an Elemental Analyser (Flash EA 1112 series, Thermo Fisher Scientific, Waltham, MA, USA) allows the combustion of solid and liquid samples at 1020 °C by means of O_2 gas. Specifically, the samples are weighed in tin capsules, with masses from 0.8 to 2 mg, according to the archaeological material to be processed, and uploaded into an autosampler. After combustion, the gases produced from the bulk samples undergo a red-ox reaction in a quartz tube, manually packed with specific reagents. In particular, we use chromium oxide, silvered cobalt oxide and reduced copper wires, separated by quartz wool. In this way, CO_2 and N_2 gases are obtained and subsequently separated by a pre-packed chromatographic column, which operates at 60 °C. Before the GC column, a filter is placed to remove water from the sample, filled with magnesium perchlorate. The EA, standing alone, also enables the measurement of carbon and nitrogen concentrations, by means of a Thermal Conductivity Detector (TCD, Thermo Fisher Scientific, Waltham, MA, USA).

For $\delta^{18}O$ analysis, the peripheral used is a Thermo Chemical Elemental Analyzer (TC-EA, Thermo Fisher Scientific, Waltham, MA, USA), where pyrolysis occurs at 1350–1450 °C, depending on the organic or inorganic origin of the sample, respectively. In this case, samples are weighed in silver capsules, with weights in the range of a few mg, but still uploaded in an autosampler, as for the EA. The reaction happens in a glassy carbon tube filled with granulate glassy carbon and positioned in a ceramic tube. The catalytic breaking of molecules at high temperature generates CO (and N_2, chromatographically separated) gas on which the oxygen isotope ratio is measured.

For compound-specific isotope analysis (CSIA), the laboratory is equipped with a Gas Chromatograph (Trace 1310- GC Isolink II, Thermo Fisher Scientific, Waltham, MA, USA), which allows the selection of specific molecules whose isotope ratios can be measured. The instrument can measure both $\delta^{13}C$ and $\delta^{15}N$, as it is equipped with a liquid nitrogen trap that retains the CO_2 produced by the combustion of the sample, allowing only N_2 to pass through. Depending on the molecule to be measured (amino acids or fatty acids), a specific chromatography column is assembled. In this case, the sample undergoes a specific chemical pretreatment (derivatisation), to enhance the volatility of the molecule to be detected [19]. The system is equipped with a 150-position autosampler.

All the peripherals are connected to the IRMS in continuous flow, by means of a CONFLO IV, which allows the passage of both the sample and reference gases, alternated according to the selected measurement method. The reference gases used are CO_2 for $\delta^{13}C$ measurements, N_2 for $\delta^{15}N$ measurements and CO for $\delta^{18}O$ one.

2.2. Measurement Methods

The entire instrumentation is managed by Isodat 3.0 Thermo Scientific software, which allows the setting up of specific measurement methods according to sample type and the isotope ratio to be measured. For example, at the iCONa lab, a method was created for measuring the combined $\delta^{15}N$ and $\delta^{13}C$ of collagen samples for paleodiet investigation. This method exploits the carbon versus nitrogen dilution method and allows also for the measurement of C and N concentrations in a single 480-s run using around 1 mg of each sample. Given the often limited and valuable nature of collagen samples, this method offers a micro-destructive approach that maximises data recovery from approximately 1 mg of each sample, alongside significant time saving.

In our laboratory measurement protocols, internationally certified standards (IAEA standards) are used to accurately calibrate raw measurements. They are included in each run list, every 24 samples, also to check the quality of the measurement and so to calculate analytical precision on precious samples that do not allow for replicate analysis. Typical analytical precision is 0.1‰ for $\delta^{13}C$ and 0.2‰ for $\delta^{15}N$.

Figure 1 shows the standardised variable Z calculated for the $\delta^{13}C$ measurements of the standard IAEA-CH-6 (sucrose, certified $\delta^{13}C = -10.449 \pm 0.033$ ‰) over three years (2020–2022), and of the standard IAEA-CH-3 (cellulose, certified $\delta^{13}C = -24.724 \pm 0.041$ ‰) for approximately one and a half years (2023- half 2024). Both the standards are used for the sample's calibration in the combined measurements of carbon and nitrogen isotopes, whereas the CH-6 standard was replaced with the CH-3 one, due to its lack of availability until recently. The Z value quantifies the distance between a data point and the mean of a dataset; it is therefore obtained by the difference between each observed value and the average, related to its own standard deviation [20].

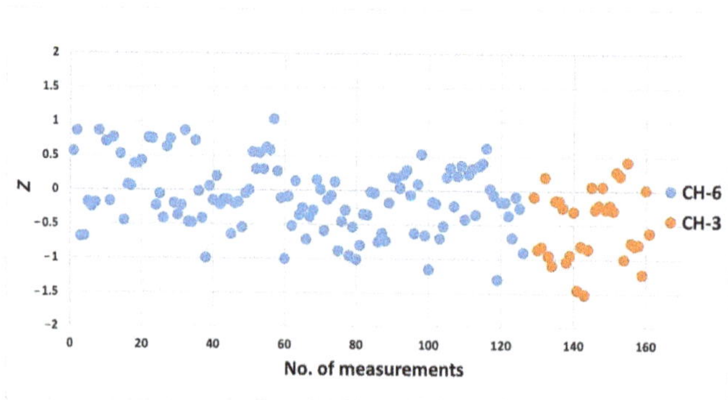

Figure 1. Variable Z values calculated for the standards IAEA-CH-6 (blue circles) and IAEA-CH-3 (orange circles).

Figure 1 indicates that all the measurements are in the range of two standard deviations (Z values less than 2) from the mean and most in that of one standard deviation (Z values less than 1).

2.3. *Overview of Samples Processed*

Figure 2a shows the quantitative distribution of the samples processed at the iCONa lab, from the installation year to date, divided into disciplinary research fields. In addition, the quantity of the standards used for measurement calibrations and quality check is presented.

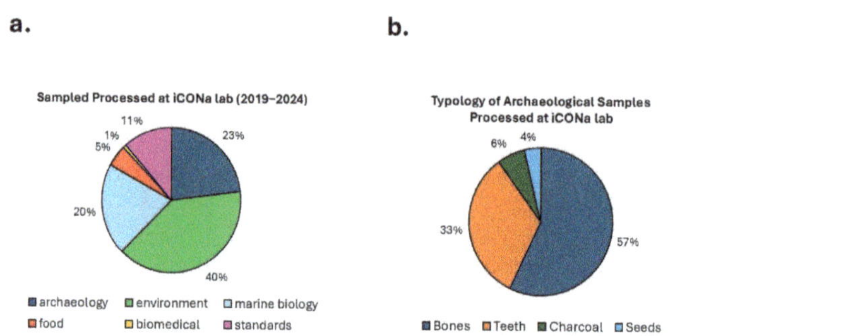

Figure 2. Pie charts representing the distribution of samples analysed at iCONa lab between 2019 and 2024 according to the disciplinary research field (**a**) and the typology of archaeological samples (**b**).

A first observation clearly shows that the environmental section prevails, specifically because it includes many different applications in this field [21–23] and thus there is diversity in the matrices analysed (wood, roots, leaves, soils, microalgal biomass, etc.). The marine biology category specifically regards analysis of different fish species, for studies of changes in trophic levels influenced by environmental and anthropogenic factors.

In addition, it can be observed that biomedical applications represent the smallest slice of the graph, as this field has recently been dealt within our study, but it is clearly expanding due to the numerous collaborations with biological and medical research teams [24].

The typology of archaeological samples processed at the iCONa lab, as presented in Figure 2b, reveals a marked emphasis on osteoarchaeological materials: 57% of the samples consist of bones, while teeth account for 33%. Conversely, archaeobotanical samples, specifically charcoal and seeds, represent a considerably smaller fraction of the laboratory's workload, at 6% and 4%, respectively. The observed variation in sample frequency directly corresponds to the broad spectrum of research questions addressed by the iCONa lab and its collaborative partners.

Each type of sample necessitates tailored pretreatment protocols to ensure reliable isotopic analyses. The following paragraph will briefly explore the specific preparation techniques needed for each archaeological material.

2.4. *Sampling and Pretreating Archaeological Remains for Isotopic Analyses: Challenges and Advantages of Micro-Destructive Techniques*

Stable isotope analysis is an inherently micro-destructive technique. However, recent advancements in isotopic methods and sample pretreatment protocols have significantly reduced the amount of archaeological material needed. Despite these improvements, careful sample selection remains essential to balance scientific goals with the preservation of valuable archaeological specimens. Adopting a research-question-driven approach is essential, ensuring that materials are only sampled when non-invasive methods cannot yield the required information. Consequently, an increasing body of research is focusing on establishing best practices for sampling archaeological organic materials for stable isotope analysis, addressing both ethical and scientific concerns associated with the irreversible nature of these procedures [11,12,15,16,25].

Different sampling strategies are required depending on the material type (e.g., archaeological bone or tooth), the specific research question (e.g., early life vs. later life dietary habits) and the type of isotopic analysis being performed (e.g., carbon and nitrogen isotope analysis for diet, or strontium isotope analysis for mobility). Sample preservation also needs to be considered, as diagenesis—chemical, physical and biological alterations occurring in the archaeological record—can impact the integrity of the material and the reliability of isotopic data [26]. Furthermore, effective pretreatment protocols are needed to avoid sample contamination, to promote methodological transparency and to ensure reliable data interpretation [17,27–29].

Table 1 provides a summary of the pretreatment protocols, required sample sizes and types of isotopic analysis performed at the iCONa lab, along with the distinct information that each methodology provides. The specified sample sizes are intended for samples in a good state of preservation; samples with poor conservation status usually require increased quantities.

Our laboratory employs a comprehensive analytical approach, integrating established traditional methodologies with advanced isotopic techniques. Notably, our expertise in incremental dentine analysis represents a key differentiator in this field; to our knowledge, our facility is currently the sole laboratory in Italy conducting this specialised methodology in-house. Additionally, our commitment to advancing analytical capabilities is demonstrated by the current development of a dedicated setup for compound-specific isotope analysis of amino acids (CSIA-AAs), a service not currently available in other Italian research facilities.

Table 1. Summary of the sample types, the required mass for each sample, the pretreatment protocols utilised, the isotopic analyses conducted at the iCONa lab and the types of information that can be obtained from these analyses.

Sample Type	Sample Weight	Pretreatment Protocol	Information
Bone	300–500 mg	Bone collagen extraction described in [29], omitting the ultrafiltration step	Paleodiet and animal management practices reconstruction ($\delta^{13}C$, $\delta^{15}N$) Radiocarbon dating (^{14}C)
Tooth	Variable	Dentine collagen extraction [30]	Paleodiet and animal management practices reconstruction ($\delta^{13}C$, $\delta^{15}N$) Radiocarbon dating (^{14}C)
	Variable	Incremental dentine sampling [31,32]	Infant feeding practices and early life dietary habits ($\delta^{13}C$, $\delta^{15}N$)
	5–10 mg	Enamel sampling and strontium extraction [33]	Spatial mobility ($^{87}Sr/^{86}Sr$)
Charcoal/Seeds	Min. 5–10 mg	Standard ABA protocol [34] or its modified versions [27], depending on the material type and preservation state	Agricultural practices, paleoenvironmental and paleoclimate reconstruction ($\delta^{13}C$, $\delta^{15}N$) Radiocarbon dating (^{14}C)

3. Stable Isotope Analysis to Reconstruct Past Subsistence Strategies with a High Level of Resolution: Recent Methodological Advances

3.1. Stable Carbon and Nitrogen Isotope Analysis: Basic Principles and Limitations

Stable isotope analysis of human bones and teeth has become an essential tool for addressing a wide range of archaeological research questions related to past subsistence strategies and human–environment interactions. The main isotopic proxies utilised to investigate past dietary practices are stable carbon ($\delta^{13}C$) and nitrogen ($\delta^{15}N$) isotopes measured on bone and dentine collagen. Bulk collagen stable carbon values allow us to estimate the proportions of C_3 plants and cereals (e.g., wheat, barley, rye, legumes, fruits, vegetables) versus C_4 cereals (e.g., millet, sorghum) in the diet, and to distinguish between terrestrial and marine dietary sources, as each has distinct carbon isotopic signatures [2,9,14]. In contrast, nitrogen isotopes offer insights into the trophic levels within the food chain and allow us to infer the protein contribution to the consumer's diet [9,14,35]. Therefore, the combination of stable carbon and nitrogen analysis is essential for comprehensive dietary reconstruction. To accurately interpret human isotopic measurements, it is also necessary to establish an isotopic baseline of both faunal and plant remains. Local baselines are essential for reconstructing agricultural practices and animal management strategies and for gaining more detailed insight into human dietary choices [36–38].

However, stable isotope analysis of human skeletal remains presents limitations that can restrain the reliability of dietary reconstructions. One primary drawback is its averaging nature, which often obscures individual variations in diet and fails to capture short-term or seasonal changes, instead reflecting only longer-term dietary patterns [39]. Additionally, the complexity of isotopic signatures—shaped by factors such as metabolic processes, baseline variability and different access to local resources—complicates result interpretation, particularly when trying to estimate the dietary contribution of different food sources that may exhibit overlapping isotopic signatures [40,41].

To address these limitations and improve the resolution of data interpretation, researchers have developed advanced methodologies over the past few decades, including incremental dentine analysis, compound-specific stable isotope analysis of amino acids (CSIA-AAs) and dietary mixing models. Incremental dentine analysis allows us to explore early life dietary changes, linking isotopic signatures to specific life stages and enabling palaeodietary reconstruction with a high temporal resolution [31,32,42,43]. CSIA-AAs enhance the precision of dietary source identification by analysing the isotopic ratios of individual amino acids, allowing for differentiation of various food sources more effectively compared to bulk collagen stable isotope analysis [44,45]. Furthermore, dietary mixing models, which can incorporate non-isotopic prior information, can quantify the relative contributions of the available food sources for past populations [41,46,47].

In the next paragraph, an overview of these methodological and interpretative advances in the isotopic field, which are also employed at the iCONa lab, will be provided. This discussion will highlight how these innovative techniques enhance insights into dietary patterns and refine isotopic-based historical narratives.

3.2. Incremental Dentine Analysis: Investigating Early Life Dietary Habits with a High Temporal Resolution

Due to different turnover rates, bones and teeth provides distinct types of information tailored to the required research question. Bones undergo continuous remodelling throughout life, influenced by factors such as biological sex, health, nutrition, age and the type of skeletal element, incorporating the isotopic signatures of the diet consumed during that period. Consequently, the turnover rate is an essential factor to consider when selecting bone samples for stable isotope analysis, depending on the research focus. For example, ribs, which remodel quickly, reflect dietary intake from the last few years of life, while femurs, which remodel more slowly, capture a longer-term dietary signal over approximately a decade [48,49].

In contrast, teeth do not remodel after they form, which allows them to preserve an in vivo isotopic record of diet from infancy through to adolescence [32,50]. Each tooth forms during a specific life stage, allowing for the study of dietary patterns at different points in an individual's early life. For instance, the first molar records dietary information from early childhood, while the third molar provides insight into diet during late adolescence [51]. By sampling multiple teeth from the same individual, it is possible to reconstruct a more detailed dietary history over time [52]. However, bulk collagen analysis of tooth dentine provides an averaged isotopic signal across the entire period of tooth formation, limiting the ability to study short-term dietary changes. Refined methods, such as sectioning dentine increments, allow for more precise insights into dietary shifts over shorter periods [31,32,42,43,53,54].

Dentine is a calcified tissue that grows sequentially from the crown of the tooth towards the root apex, forming a series of incremental layers. In humans, the daily secretion rate of dentine ranges from approximately 2 to 6 micrometres (μm) [55]. By sequentially sampling and performing stable carbon and nitrogen isotope analysis on each dentine increment, high temporal resolution insights into past individuals' dietary habits can be achieved [54]. Over the past 20 years, multiple sectioning techniques have been developed [31,32,42,43,53–55]. At the iCONa lab, we employ both the micro-punch method developed by Czermak et al. [32] and the micro-slice technique introduced by Beaumont et al. [31] for sequential sampling of human dentine. Incremental dentine analysis is particularly useful for reconstructing breastfeeding and weaning practices in past populations using measurements of stable carbon ($\delta^{13}C$) and nitrogen ($\delta^{15}N$) isotope ratios on dentine increments from the first permanent molar, which records isotopic signals from approximately three months postpartum to around nine and half years of age [52,56]. During exclusive breastfeeding, infants consume ^{15}N-enriched milk, placing them at a higher trophic level in the food chain compared to their mothers [57]. A decrease in $\delta^{15}N$ identifies the transition from exclusive breastfeeding to the introduction of complementary foods (e.g., cereals, fruits), which are lower in protein compared to human milk [58]. The cessation of the decrease in $\delta^{15}N$ values, followed by an isotopic plateau, is employed to estimate the end of breastfeeding [57,59].

Consequently, incremental dentine analysis not only enhances our understanding of past dietary habits at a high temporal resolution but also offers detailed insights into the sociocultural dynamics of infant feeding strategies in ancient populations [57].

3.3. *Compound-Specific Stable Isotope Analysis of Amino Acids: Overcoming the Limitations of Bulk Stable Isotope Analysis*

The application of compound-specific stable isotope analysis of amino acids (CSIA-AAs) has emerged as a transformative approach for reconstructing ancient diets, offering enhanced resolution compared to bulk isotope methods.

By analysing the isotopic signatures of individual amino acids, CSIA enables the disentanglement of complex dietary inputs and metabolic pathways, providing critical insights into past ecological interactions. For nitrogen isotopes, the ability to distinguish between trophic and source amino acids allows for precise determination of trophic positions, revealing the balance between plant and animal protein consumption in human and animal diets [60]. Similarly, carbon isotope analysis of essential amino acids facilitates the tracing of primary producers within food webs, enabling the differentiation of terrestrial and aquatic dietary sources or the identification of C_3 versus C_4 plant contributions [61]. This specificity is particularly valuable in archaeological contexts where mixed diets, isotopic mixing zones or environmental variability can obscure signals in bulk isotope data [62].

Furthermore, CSIA's capability to target individual biochemical compounds reduces the impact of diagenetic alterations, which can significantly affect the reliability of bulk analyses [19]. Advances in preparation techniques and analytical sensitivity have also minimised sample size requirements, making CSIA suitable for rare or fragmented archaeological specimens [45,63].

This methodology not only reconstructs dietary practices with a requirement for smaller sample quantities and reduced susceptibility to contamination but also sheds light on broader ecological and environmental conditions, contributing to a holistic understanding of ancient lifeways and their adaptation to changing environments [60,64].

3.4. *Bayesian Dietary Mixing Models: Estimating Quantitative Dietary Reconstructions*

Bayesian dietary mixing models represent a cutting-edge approach for quantitatively reconstructing individual diets by analysing the relative contributions of various food sources through isotopic data from consumers and their dietary sources [41,46,47,65]. In the field of archaeology, stable carbon and nitrogen isotopic ratios measured on bone collagen serve as the primary isotopic proxies for reconstructing past diets; however, other isotopic ratios and tissue types can also be utilised for this purpose [41,66].

Accurate dietary reconstruction through dietary mixing models requires careful consideration of several key factors. One significant challenge is the isotopic fractionation that occurs during metabolic processes, resulting in diet-to-tissue isotopic offsets [67,68]. These offsets, which accounts for the isotopic differences between ingested foods and the consumer tissues, are typically estimated through controlled feeding experiments [46,47,69].

Another consideration in quantitative dietary reconstructions is the estimation of the isotopic composition of macronutrients—specifically, proteins compared to lipids and carbohydrates—in food remains. Potential variations between the isotopic values obtained from food remains (e.g., faunal bone collagen) and those of the edible parts of the food (e.g., meat protein) require the use of offset corrections [47,69]. Additionally, when dealing with plant and cereal remains, it is essential to account for charring effects that may have altered their in vivo isotopic signatures [46,47].

It is also necessary to consider the macronutrient composition of the food groups available for the studied population [69]. Animal-based foods generally contain higher protein levels, while plant-based foods are predominantly carbohydrate-rich [70]. Moreover, macronutrient composition can vary significantly across different animal species [69]. By considering the average macronutrient profiles of each food group, and their relative uncertainties, dietary mixing models can more accurately reflect the dietary contributions of food sources [47].

Lastly, it is essential to acknowledge dietary routing mechanisms by which carbon and nitrogen from dietary macronutrients (proteins, carbohydrates and lipids) are incorporated into bone collagen [71]. Controlled feeding studies have shown that nitrogen isotopes are sourced from dietary protein, while carbon isotopes predominantly originate from protein, with lesser contributions from carbohydrates and lipids [46,47,71,72].

Advanced models, such as ReSources (previously FRUITS; [46]), effectively address these parameters and their associated uncertainties. Notably, ReSources allows for the integration of non-isotopic prior constraints, which can be derived from historical records, zooarchaeological findings and archaeobotanical data [73,74]. This incorporation significantly enhances the precision of dietary estimates, especially when it comes to distinguishing between food sources that exhibit overlapping isotopic signatures [46,47].

Bayesian dietary mixing models therefore represent a significant advancement in stable isotope analysis, offering a quantitative approach to reconstructing past diets. By incorporating multiple parameters, including diet-to-tissue offsets, macronutrient and isotopic composition of food groups, dietary routing and the possibility to integrate non-isotopic prior information, these models provide more detailed insights into past human subsistence strategies [40,46,69,75,76].

However, it is essential to also acknowledge the limitations of dietary mixing models. The reliability of dietary reconstructions heavily relies on the quality of the input data, including isotopic values for potential food sources to reconstruct local baselines [39,41,70]. Furthermore, while quantitatively reconstructing paleodiets, assumptions of the linearity of isotopic fractionation and the homogeneity of food sources within the selected population may not always accurately reflect the complexities of the human diet [70].

4. Research Lines and Collaborations of iCONa Lab in the Archaeological Field

At the iCONa lab we explore a range of research questions related to past human subsistence strategies, mobility, agricultural and animal management practices across different historical periods and geographic areas, in collaboration with academic and cultural institutions.

A key focus of our research is isotope-based past dietary reconstruction, which we explore with different levels of resolution. One significant line of research delves into the exploration of socioeconomic practices of the Basque region (Spain) during the Middle Ages. Through stable carbon and nitrogen isotope analyses of human, animal and plant remains, we reconstructed the dietary patterns as well as agricultural and animal husbandry techniques utilised by medieval rural communities within this region. This work has been conducted in the framework of an ongoing collaboration with the University of the Basque Country [77,78].

Another research pathway aims at investigating palaeodietary practices with a higher level of resolution, by employing advanced isotopic techniques. We utilised incremental dentine stable carbon and nitrogen isotope analysis to gain detailed insights into breastfeeding and weaning practices across the Roman Empire, as part of a collaborative effort with the Max Planck Institute of Geoanthropology, the Archaeological Park of Pompeii and the Archaeological Park of Ostia Antica. This study inferred a possible link between infant feeding strategies and settlement complexity across the Roman Empire. The results suggested shorter breastfeeding durations in highly complex urban centres like Pompeii (79 CE), compared to rural sites like Ostia AVM (1st–2nd centuries CE), potentially influenced by factors such as access to medical knowledge and socioeconomic conditions [57].

Our collaboration with the Archaeological Park of Ostia Antica also included the investigation of subsistence practices of the Late-Antiquity individuals buried in the Antemurale Area of *Portus Romae*. The employment of a multidisciplinary approach allowed for the reconstruction of the cultural and biological traits of this population. Stable carbon and nitrogen isotope analyses suggested a diet primarily based on C_3 terrestrial resources, while strontium isotope analysis implied their local geographic origins [79].

In pursuit of exploring past human behaviours with multi-scale approaches, one of our research efforts focuses on the collection of extensive isotopic data across various chronological and geographical contexts, particularly emphasising Europe and the Mediterranean Basin. Within the framework of the Mediterranean Bioarchaeological Research Advances (MAReA) centre's efforts, several isotopic databases have been published in Open Access, serving as valuable resources for bioarchaeological research. These include CIMA: *Compendium Isotoporum Medii Aevi* [74,80], IsoMedIta: A Stable Isotope Database for Medieval Italy [81], Isotòpia: A Stable Isotope Database for Classical Antiquity [82] and MAIA: Mediterranean Archive of Isotopic dAta [83]. These databases serve as comprehensive isotopic archives that help identify research gaps and investigate a variety of archaeological questions using multi-scale approaches. These activities are framed within the Pandora & IsoMemo initiative "https://isomemo.gea.mpg.de/ (accessed on 22 July 2025)".

Our ongoing collaboration with the University of Foggia and the University of Salento is contributing to a broader understanding of dietary practices, resource management, past mobility and environmental adaptation in medieval communities from the Apulia region (Southern Italy). A collaborative effort, led by the Max Planck Institute of Geoanthropology, highlighted a farming economy based on cereal production complemented by intensive pig and ovicaprid husbandry in Late Medieval Capitanata. Dietary differences reflected a socioeconomic hierarchy, with elites likely to have greater access to animal protein and higher-quality grains [73].

While palaeodietary studies provide valuable insights into subsistence strategies of ancient populations, radiocarbon dating serves as a useful tool for establishing chronological frameworks within archaeological studies. Radiocarbon dating performed on mortars and charcoals was employed to investigate and reassess the chronologies of the encastellation phenomenon (*incastellamento*) in medieval Italy. A multidisciplinary approach was employed to establish more precise timelines of and provide novel insights into the formation and transformation processes of fortified structures between the 11th and 12th centuries CE, thereby enhancing our understanding of the historical context and significance of encastellation in northern Italy. This line of research has been carried out within the framework of PRIN 2020 Project "The Times of CASTLES: Multidisciplinary Research for a New Chronology of the Building Sites of Incastellamento (XI–XII centuries) (MIUR protocol: 20203YX58R, CUP: B53C2100029000, PI: Prof. Giovanna Bianchi).

By employing a combined approach of stable carbon and nitrogen isotope analyses alongside radiocarbon dating of bone collagen, we also investigated the individuals buried in the Selvicciola Necropolis, a notable Copper Age burial site in Central Italy. This integrative methodology revealed a burial chronology spanning approximately 2000 years and provided valuable insights into the dietary practices and burial patterns of this community [84].

Our laboratory's publication record showcases the breadth of research questions we address, employing diverse levels of resolution. These investigations collectively encompass multiple archaeological contexts, with multidisciplinary approaches made possible by our engagement with numerous institutions and collaborative partners.

5. Conclusions

This paper has emphasised the contributions, in the context of the iCONa lab, of stable isotope analysis to the field of cultural heritage, particularly focusing on archaeological remains such as bones, teeth and charred botanical materials. These analytical techniques provide valuable insights into ancient subsistence strategies, past mobility and agricultural and husbandry practices, allowing us to explore how past populations adapted to environmental, political and socioeconomic changes.

The research undertaken herein underscores the importance of acknowledging and addressing the inherent challenges associated with isotopic analyses. Key issues such as the representativeness of sampling and the necessity for methodological standardisation must be carefully examined to uphold the validity and integrity of our results. Tackling these challenges is essential for enhancing the reliability of data interpretation. Furthermore, the ongoing advancement of isotopic methodologies has proven essential in refining our understanding of past subsistence strategies with heightened resolution level.

The iCONa lab distinguishes itself from facilities operating primarily as service laboratories through its dual commitment to methodological advancement and active engagement in original research. This approach, enhanced by our collaborative efforts with a variety of academic and institutional partners, ensures that our analytical advances translate into contributions which address multiple research questions in the cultural heritage field.

Author Contributions: Conceptualisation, N.M., S.A. and C.L., methodology, N.M., S.A., M.R.d.C. and V.G.; software, N.M., S.A., M.R.d.C. and V.G.; validation, N.M., S.A., M.R.d.C., V.G. and C.L.; formal analysis, N.M. and S.A.; investigation, N.M. and S.A.; resources, C.L.; data curation, N.M., S.A., M.R.d.C. and V.G.; writing—original draft preparation, N.M., S.A., M.R.d.C. and V.G.; writing—review and editing, N.M., S.A., M.R.d.C., V.G. and C.L.; visualisation, N.M., S.A., M.R.d.C. and V.G.; supervision, C.L.; project administration, C.L.; funding acquisition, C.L. All authors have read and agreed to the published version of the manuscript.

Funding: This research received no external funding.

Data Availability Statement: The original contributions presented in this study are included in the article. Further inquiries can be directed to the corresponding authors.

Conflicts of Interest: The authors declare no conflicts of interest.

References

1. Cotrufo, F.; Pressler, Y. *A Primer on Stable Isotopes in Ecology*; Oxford University Press: Oxford, UK, 2023; ISBN 0-19-885449-8.
2. Somerville, A.D.; Beasley, M.M. Exploring Human Behavior Through Isotopic Analyses: Tools, Scales, and Questions. In *Exploring Human Behavior Through Isotope Analysis*; Interdisciplinary Contributions to Archaeology; Beasley, M.M., Somerville, A.D., Eds.; Springer International Publishing: Cham, Switzerland, 2023; pp. 9–32, ISBN 978-3-031-32266-2.
3. Kelly, S.D. Using Stable Isotope Ratio Mass Spectrometry (IRMS). In *Food Authentication and Traceability*; Woodhead Publishing: Cambridge, UK, 2003. [CrossRef]
4. Giannioti, Z.; Ogrinc, N.; Suman, M.; Camin, F.; Bontempo, L. Isotope Ratio Mass Spectrometry (IRMS) Methods for Distinguishing Organic from Conventional Food Products: A Review. *TrAC Trends Anal. Chem.* **2024**, *170*, 117476. [CrossRef]

5. Tea, I.; De Luca, A.; Schiphorst, A.-M.; Grand, M.; Barillé-Nion, S.; Mirallié, E.; Drui, D.; Krempf, M.; Hankard, R.; Tcherkez, G. Stable Isotope Abundance and Fractionation in Human Diseases. *Metabolites* **2021**, *11*, 370. [CrossRef] [PubMed]
6. Roncone, A.; Kelly, S.D.; Giannioti, Z.; Hauk, C.; Caillet, C.; Newton, P.N.; Perez-Mon, C.; Bontempo, L. Stable Isotope Ratio Analysis: An Emerging Tool to Trace the Origin of Falsified Medicines. *TrAC Trends Anal. Chem.* **2024**, *174*, 117666. [CrossRef]
7. Meier-Augenstein, W. *Stable Isotope Forensics: Methods and Forensic Applications of Stable Isotope Analysis*; John Wiley & Sons: Hoboken, NJ, USA, 2017; ISBN 978-1-119-08020-6.
8. Criss, R.E. *Principles of Stable Isotope Distribution*; Oxford University Press: Oxford, UK, 1999; ISBN 978-0-19-511775-2.
9. Katzenberg, M.A.; Waters-Rist, A.L. Stable Isotope Analysis: A Tool for Studying Past Diet, Demography, and Life History. In *Biological Anthropology of the Human Skeleton*; Katzenberg, M.A., Grauer, A.L., Eds.; Wiley: Hoboken, NJ, USA, 2018; pp. 467–504, ISBN 978-1-119-15161-6.
10. Mook, W.G. Introduction to Isotope Hydrology. In *Stable and Radioactive Isotopes of Hydrogen, Oxygen and Carbon*; Taylor & Francis Group: London, UK, 2005; ISBN 978-0-415-39805-3.
11. Mays, S.; Elders, J.; Humphrey, L.T.; White, W.; Marshall, P. *Science and the Dead: A Guideline for the Destructive Sampling of Archaeological Human Remains for Scientific Analysis*; English Heritage with the Advisory Panel on the Archaeology of Burials in England: England, UK, 2013.
12. Pálsdóttir, A.H.; Bläuer, A.; Rannamäe, E.; Boessenkool, S.; Hallsson, J.H. Not a Limitless Resource: Ethics and Guidelines for Destructive Sampling of Archaeofaunal Remains. *R. Soc. Open Sci.* **2019**, *6*, 191059. [CrossRef] [PubMed]
13. Stantis, C.; Schaefer, B.J.; Correia, M.A.; Alaica, A.K.; Huffer, D.; Plomp, E.; Di Giusto, M.; Chidimuro, B.; Rose, A.K.; Nayak, A.; et al. Ethics and Applications of Isotope Analysis in Archaeology. *Am. J. Biol. Anthropol.* **2024**, *186*, e24992. [CrossRef]
14. Lee-Thorp, J.A. On isotopes and old bones. *Archaeometry* **2008**, *50*, 925–950. [CrossRef]
15. Mitchell, P.D.; Brickley, M. *Updated Guidelines to the Standards for Recording Human Remains*; Chartered Institute for Archaeologists: Earley, UK, 2017; ISBN 978-0-948393-27-3.
16. Vaiglova, P. Sampling Archaeological Remains for Isotopic Analyses. In *Field Sampling for Laboratory Analyses in Archaeology*; The Cyprus Institute: Nicosia, Cyprus, 2023; pp. 53–64, ISBN 978-9925-7759-4-1.
17. Roberts, P.; Fernandes, R.; Craig, O.E.; Larsen, T.; Lucquin, A.; Swift, J.; Zech, J. Calling All Archaeologists: Guidelines for Terminology, Methodology, Data Handling, and Reporting When Undertaking and Reviewing Stable Isotope Applications in Archaeology. *Rapid Comm. Mass Spectrom.* **2018**, *32*, 361–372. [CrossRef]
18. Ricci, P.; Sirignano, C.; Altieri, S.; Pistillo, M.; Santoriello, A.; Lubritto, C. Paestum Dietary Habits during the Imperial Period: Archaeological Records and Stable Isotope Measurement. *Acta IMEKO* **2016**, *5*, 26–32. [CrossRef]
19. Corr, L.T.; Berstan, R.; Evershed, R.P. Optimisation of Derivatisation Procedures for the Determination of δ^{13} C Values of Amino Acids by Gas Chromatography/Combustion/Isotope Ratio Mass Spectrometry. *Rapid Comm. Mass Spectrom.* **2007**, *21*, 3759–3771. [CrossRef]
20. James, G.; Witten, D.; Hastie, T.; Tibshirani, R. *An Introduction to Statistical Learning*; Springer Texts in Statistics; Springer: New York, NY, USA, 2013; Volume 103, ISBN 978-1-4614-7137-0.
21. Altieri, S.; Niccoli, F.; Kabala, J.P.; Liyaqat, I.; Battipaglia, G. Influence of Drought and Minimum Temperature on Tree Growth and Water Use Efficiency of Mediterranean Species. *Dendrochronologia* **2024**, *83*, 126162. [CrossRef]
22. Niccoli, F.; Altieri, S.; Kabala, J.P.; Battipaglia, G. Fire Affects Tree Growth, Water Use Efficiency and Carbon Sequestration Ecosystem Service of Pinus Nigra Arnold: A Combined Satellite and Ground-Based Study in Central Italy. *Forests* **2023**, *14*, 2033. [CrossRef]
23. di Cicco, M.R.; Altieri, S.; Spagnuolo, A.; Ciniglia, C.; Germinario, C.; Bove, S.; Masiello, A.; Vetromile, C.; Galante, I.; Lubritto, C. Sustainable Cultivation of Galdieria Phlegrea in an IoT-Integrated Twin-Layer Photobioreactor: System Design, Growth Dynamics, and Isotopic Perspective. *Appl. Sci.* **2025**, *15*, 5220. [CrossRef]
24. Marfella, R.; Prattichizzo, F.; Sardu, C.; Fulgenzi, G.; Graciotti, L.; Spadoni, T.; D'Onofrio, N.; Scisciola, L.; La Grotta, R.; Frigé, C.; et al. Microplastics and Nanoplastics in Atheromas and Cardiovascular Events. *N. Engl. J. Med.* **2024**, *390*, 900–910. [CrossRef] [PubMed]
25. McKinley, B.M.; Brickley, M. Guidelines to the Standards for Recording Human Remains (IFA Paper No. 7). In *Institute of Field Archaeologists and British Association for Biological Anthropology and Osteoarchaeology*; University of Southampton: Southampton, UK, 2004; ISBN 0948 393 88 2.
26. Kendall, C.; Eriksen, A.M.H.; Kontopoulos, I.; Collins, M.J.; Turner-Walker, G. Diagenesis of Archaeological Bone and Tooth. *Palaeogeogr. Palaeoclimatol. Palaeoecol.* **2018**, *491*, 21–37. [CrossRef]
27. Vaiglova, P.; Snoeck, C.; Nitsch, E.; Bogaard, A.; Lee-Thorp, J. Impact of Contamination and Pre-treatment on Stable Carbon and Nitrogen Isotopic Composition of Charred Plant Remains. *Rapid Comm. Mass Spectrom.* **2014**, *28*, 2497–2510. [CrossRef]
28. Brinkkemper, O.; Braadbaart, F.; Van Os, B.; Van Hoesel, A.; Van Brussel, A.A.N.; Fernandes, R. Effectiveness of Different Pre-treatments in Recovering Pre-burial Isotopic Ratios of Charred Plants. *Rapid Comm. Mass Spectrom.* **2018**, *32*, 251–261. [CrossRef]

29. Talamo, S.; Fewlass, H.; Maria, R.; Jaouen, K. "Here We Go Again": The Inspection of Collagen Extraction Protocols for[14] C Dating and Palaeodietary Analysis. *STAR Sci. Technol. Archaeol. Res.* **2021**, *7*, 62–77. [CrossRef]
30. Richards, M.P.; Mays, S.; Fuller, B.T. Stable Carbon and Nitrogen Isotope Values of Bone and Teeth Reflect Weaning Age at the Medieval Wharram Percy Site, Yorkshire, UK. *Am. J. Phys. Anthr.* **2002**, *119*, 205–210. [CrossRef]
31. Beaumont, J.; Gledhill, A.; Lee-Thorp, J.; Montgomery, J. Childhood diet: A closer examination of the evidence from dental tissues using stable isotope analysis of incremental human dentine*. *Archaeometry* **2013**, *55*, 277–295. [CrossRef]
32. Czermak, A.; Fernández-Crespo, T.; Ditchfield, P.W.; Lee-Thorp, J.A. A Guide for an Anatomically Sensitive Dentine Microsampling and Age-alignment Approach for Human Teeth Isotopic Sequences. *Am. J. Phys. Anthr.* **2020**, *173*, 776–783. [CrossRef]
33. Lugli, F.; Cipriani, A.; Arnaud, J.; Arzarello, M.; Peretto, C.; Benazzi, S. Suspected Limited Mobility of a Middle Pleistocene Woman from Southern Italy: Strontium Isotopes of a Human Deciduous Tooth. *Sci. Rep.* **2017**, *7*, 8615. [CrossRef]
34. Fraser, R.A.; Bogaard, A.; Schäfer, M.; Arbogast, R.; Heaton, T.H.E. Integrating Botanical, Faunal and Human Stable Carbon and Nitrogen Isotope Values to Reconstruct Land Use and Palaeodiet at LBK Vaihingen an Der Enz, Baden-Württemberg. *World Archaeol.* **2013**, *45*, 492–517. [CrossRef]
35. Schoeninger, M.J.; DeNiro, M.J. Nitrogen and Carbon Isotopic Composition of Bone Collagen from Marine and Terrestrial Animals. *Geochim. Cosmochim. Acta* **1984**, *48*, 625–639. [CrossRef]
36. Fuller, B.T.; De Cupere, B.; Marinova, E.; Van Neer, W.; Waelkens, M.; Richards, M.P. Isotopic Reconstruction of Human Diet and Animal Husbandry Practices during the Classical-Hellenistic, Imperial, and Byzantine Periods at Sagalassos, Turkey. *Am. J. Phys. Anthr.* **2012**, *149*, 157–171. [CrossRef] [PubMed]
37. Bownes, J.; Clarke, L.; Buckberry, J. The Importance of Animal Baselines: Using Isotope Analysis to Compare Diet in a British Medieval Hospital and Lay Population. *J. Archaeol. Sci. Rep.* **2018**, *17*, 103–110. [CrossRef]
38. Garcia Collado, M.I. Social Archaeology of Food in Early Medieval Rural Iberia (5th–9th c. AD). PhD Thesis, Universidad del País Vasco-Euskal Herriko Unibertsitatea, Santsoena, Spain, 2020.
39. Cheung, C.; Szpak, P. Interpreting Past Human Diets Using Stable Isotope Mixing Models—Best Practices for Data Acquisition. *J. Archaeol. Method Theory* **2022**, *29*, 138–161. [CrossRef]
40. Bownes, J.M.; Ascough, P.L.; Cook, G.T.; Murray, I.; Bonsall, C. Using Stable Isotopes and a Bayesian Mixing Model (FRUITS) to Investigate Diet at the Early Neolithic Site of Carding Mill Bay, Scotland. *Radiocarbon* **2017**, *59*, 1275–1294. [CrossRef]
41. Cheung, C.; Szpak, P. Interpreting Past Human Diets Using Stable Isotope Mixing Models. *J. Archaeol. Method Theory* **2021**, *28*, 1106–1142. [CrossRef]
42. Fuller, B.T.; Richards, M.P.; Mays, S.A. Stable Carbon and Nitrogen Isotope Variations in Tooth Dentine Serial Sections from Wharram Percy. *J. Archaeol. Sci.* **2003**, *30*, 1673–1684. [CrossRef]
43. Beaumont, J.; Gledhill, A.; Montgomery, J. Isotope Analysis of Incremental Human Dentine: Towards Higher Temporal Resolution. *Bull. Int. Assoc. Paleodont.* **2014**, *8*, 212–223.
44. Ma, Y.; Grimes, V.; Van Biesen, G.; Shi, L.; Chen, K.; Mannino, M.A.; Fuller, B.T. Aminoisoscapes and Palaeodiet Reconstruction: New Perspectives on Millet-Based Diets in China Using Amino Acid δ13C Values. *J. Archaeol. Sci.* **2021**, *125*, 105289. [CrossRef]
45. Soncin, S.; Talbot, H.M.; Fernandes, R.; Harris, A.; Von Tersch, M.; Robson, H.K.; Bakker, J.K.; Richter, K.K.; Alexander, M.; Ellis, S.; et al. High-Resolution Dietary Reconstruction of Victims of the 79 CE Vesuvius Eruption at Herculaneum by Compound-Specific Isotope Analysis. *Sci. Adv.* **2021**, *7*, eabg5791. [CrossRef]
46. Fernandes, R.; Millard, A.R.; Brabec, M.; Nadeau, M.-J.; Grootes, P. Food Reconstruction Using Isotopic Transferred Signals (FRUITS): A Bayesian Model for Diet Reconstruction. *PLoS ONE* **2014**, *9*, e87436. [CrossRef]
47. Fernandes, R.; Grootes, P.; Nadeau, M.; Nehlich, O. Quantitative Diet Reconstruction of a Neolithic Population Using a Bayesian Mixing Model (FRUITS): The Case Study of Ostorf (Germany). *Am. J. Phys. Anthr.* **2015**, *158*, 325–340. [CrossRef]
48. Hedges, R.E.M.; Clement, J.G.; Thomas, C.D.L.; O'Connell, T.C. Collagen Turnover in the Adult Femoral Mid-shaft: Modeled from Anthropogenic Radiocarbon Tracer Measurements. *Am. J. Phys. Anthr.* **2007**, *133*, 808–816. [CrossRef] [PubMed]
49. Fahy, G.E.; Deter, C.; Pitfield, R.; Miszkiewicz, J.J.; Mahoney, P. Bone Deep: Variation in Stable Isotope Ratios and Histomorphometric Measurements of Bone Remodelling within Adult Humans. *J. Archaeol. Sci.* **2017**, *87*, 10–16. [CrossRef]
50. Turner-Walker, G. The Chemical and Microbial Degradation of Bones and Teeth. *Adv. Hum. Palaeopathol.* **2008**, *592*, 3–29. [CrossRef]
51. AlQahtani, S.J.; Hector, M.P.; Liversidge, H.M. Brief Communication: The London Atlas of Human Tooth Development and Eruption. *Am. J. Phys. Anthr.* **2010**, *142*, 481–490. [CrossRef]
52. Cocozza, C.; Fernandes, R. Amalthea: A Database of Isotopic Measurements on Archaeological and Forensic Tooth Dentine Increments. *J. Open Archaeol. Data* **2021**, *9*, 1–7. [CrossRef]
53. Eerkens, J.W.; Berget, A.G.; Bartelink, E.J. Estimating Weaning and Early Childhood Diet from Serial Micro-Samples of Dentin Collagen. *J. Archaeol. Sci.* **2011**, *38*, 3101–3111. [CrossRef]
54. Beaumont, J.; Montgomery, J. Oral Histories: A Simple Method of Assigning Chronological Age to Isotopic Values from Human Dentine Collagen. *Ann. Hum. Biol.* **2015**, *42*, 407–414. [CrossRef]

55. Czermak, A. Diet Reconstruction Based on C/N Stable Isotope Analysis: What Can It Contribute to Address Questions on Cultural Change. In *Archaeology, History and Biosciences*; Brather-Walter, S., Ed.; De Gruyter: Berlin, Germany, 2019; pp. 181–198, ISBN 978-3-11-061665-1.
56. Reynard, L.M.; Tuross, N. The Known, the Unknown and the Unknowable: Weaning Times from Archaeological Bones Using Nitrogen Isotope Ratios. *J. Archaeol. Sci.* **2015**, *53*, 618–625. [CrossRef]
57. Cocozza, C.; Harris, A.J.; Formichella, G.; Pedrucci, G.; Rossi, P.F.; D'Alessio, A.; Amoretti, V.; Zuchtriegel, G.; O'Reilly, M.; Mantile, N. High-Resolution Isotopic Data Link Settlement Complexification to Infant Diets within the Roman Empire. *PNAS Nexus* **2025**, *4*, pgae566. [CrossRef]
58. Schurr, M.R. Using Stable Nitrogen-isotopes to Study Weaning Behavior in Past Populations. *World Archaeol.* **1998**, *30*, 327–342. [CrossRef]
59. Cocozza, C.; Fernandes, R.; Ughi, A.; Groß, M.; Alexander, M.M. Investigating Infant Feeding Strategies at Roman Bainesse through Bayesian Modelling of Incremental Dentine Isotopic Data. *Intl. J. Osteoarchaeol.* **2021**, *31*, 429–439. [CrossRef]
60. McMahon, K.W.; Polito, M.J.; Abel, S.; McCarthy, M.D.; Thorrold, S.R. Carbon and Nitrogen Isotope Fractionation of Amino Acids in an Avian Marine Predator, the Gentoo Penguin (*Pygoscelis papua*). *Ecol. Evol.* **2015**, *5*, 1278–1290. [CrossRef] [PubMed]
61. O'Connell, T.C. 'Trophic' and 'Source' Amino Acids in Trophic Estimation: A Likely Metabolic Explanation. *Oecologia* **2017**, *184*, 317–326. [CrossRef] [PubMed]
62. Fogel, M.L.; Tuross, N. Transformation of Plant Biochemicals to Geological Macromolecules during Early Diagenesis. *Oecologia* **1999**, *120*, 336–346. [CrossRef] [PubMed]
63. Evershed, R.P. ORGANIC RESIDUE ANALYSIS IN ARCHAEOLOGY: THE ARCHAEOLOGICAL BIOMARKER REVOLUTION*. *Archaeometry* **2008**, *50*, 895–924. [CrossRef]
64. Vokhshoori, N.L.; Rick, T.C.; Braje, T.J.; McCarthy, M.D. Preservation of Stable Isotope Signatures of Amino Acids in Diagenetically Altered Middle to Late Holocene Archaeological Mollusc Shells. *Geochim. Cosmochim. Acta* **2023**, *352*, 36–50. [CrossRef]
65. Hart, J.P. Human and Dog Bayesian Dietary Mixing Models Using Bone Collagen Stable Isotope Ratios from Ancestral Iroquoian Sites in Southern Ontario. *Sci. Rep.* **2023**, *13*, 7177. [CrossRef] [PubMed]
66. Szpak, P.; Metcalfe, J.Z.; Macdonald, R.A. Best Practices for Calibrating and Reporting Stable Isotope Measurements in Archaeology. *J. Archaeol. Sci. Rep.* **2017**, *13*, 609–616. [CrossRef]
67. Remien, C.H. Modeling the Dynamics of Stable Isotope Tissue-Diet Enrichment. *J. Theor. Biol.* **2015**, *367*, 14–20. [CrossRef]
68. Kadye, W.T.; Redelinghuys, S.; Parnell, A.C.; Booth, A.J. Exploring Source Differences on Diet-Tissue Discrimination Factors in the Analysis of Stable Isotope Mixing Models. *Sci. Rep.* **2020**, *10*, 15816. [CrossRef]
69. Fernandes, R. A Simple(R) Model to Predict the Source of Dietary Carbon in Individual Consumers: A Simple(r) Model to Predict the Source of Dietary Carbon. *Archaeometry* **2016**, *58*, 500–512. [CrossRef]
70. Lewis, M.C.; Sealy, J.C. Coastal Complexity: Ancient Human Diets Inferred from Bayesian Stable Isotope Mixing Models and a Primate Analogue. *PLoS ONE* **2018**, *13*, e0209411. [CrossRef]
71. Froehle, A.W.; Kellner, C.M.; Schoeninger, M.J. FOCUS: Effect of Diet and Protein Source on Carbon Stable Isotope Ratios in Collagen: Follow up to Warinner and Tuross (2009). *J. Archaeol. Sci.* **2010**, *37*, 2662–2670. [CrossRef]
72. Fernandes, R.; Nadeau, M.-J.; Grootes, P.M. Macronutrient-Based Model for Dietary Carbon Routing in Bone Collagen and Bioapatite. *Archaeol. Anthr. Sci.* **2012**, *4*, 291–301. [CrossRef]
73. Cocozza, C.; Teegen, W.-R.; Vigliarolo, I.; Favia, P.; Giuliani, R.; Muntoni, I.M.; Oione, D.; Clemens, L.; Groß, M.; Roberts, P. A Bayesian Multi-Proxy Contribution to the Socioeconomic, Political, and Cultural History of Late Medieval Capitanata (Southern Italy). *Sci. Rep.* **2023**, *13*, 4078. [CrossRef]
74. Cocozza, C.; Cirelli, E.; Groß, M.; Teegen, W.-R.; Fernandes, R. Presenting the Compendium Isotoporum Medii Aevi, a Multi-Isotope Database for Medieval Europe. *Sci. Data* **2022**, *9*, 354. [CrossRef] [PubMed]
75. Moore, J.W.; Semmens, B.X. Incorporating Uncertainty and Prior Information into Stable Isotope Mixing Models. *Ecol. Lett.* **2008**, *11*, 470–480. [CrossRef]
76. Phillips, D.L.; Inger, R.; Bearhop, S.; Jackson, A.L.; Moore, J.W.; Parnell, A.C.; Semmens, B.X.; Ward, E.J. Best Practices for Use of Stable Isotope Mixing Models in Food-Web Studies. *Can. J. Zool.* **2014**, *92*, 823–835. [CrossRef]
77. García-Collado, M.I.; Ricci, P.; Catalán Ramos, R.; Altieri, S.; Lubritto, C.; Quirós Castillo, J.A. Palaeodietary Reconstruction as an Alternative Approach to Poorly Preserved Early Medieval Human Bone Assemblages: The Case of Boadilla (Toledo, Spain). *Archaeol. Anthr. Sci.* **2019**, *11*, 3765–3782. [CrossRef]
78. García-Collado, M.I.; Quirós Castillo, J.A.; Tereso, J.P.; Seabra, L.; Lubritto, C.; Altieri, S.; Ricci, P. First Direct Evidence of Agrarian Practices in the Alava Plateau (Northern Iberia) During the Middle Ages Through Carbon and Nitrogen Stable Isotope Analyses of Charred Seeds. *Environ. Archaeol.* **2022**, 1–11. [CrossRef]
79. De Angelis, F.; Vaccaro, S.; Romboni, M.; Di Cicco, M.R.; Mantile, N.; Altieri, S.; Mezzogiorno, A.; Blundo, M.L.; Rickards, O.; Lubritto, C. Echoes from the Past: Bioarchaeological Insights into the Burial Grounds of Portus Romae. *J. Archaeol. Sci. Rep.* **2025**, *61*, 104931. [CrossRef]

80. Cocozza, C.; Cirelli, E.; Groß, M.; Teegen, W.R.; Fernandes, R. *Compendium Isotoporum Medii Aevi (CIMA)*; Pandora: Copenhagen, Denmark, 2021. [CrossRef]
81. Mantile, N.; Fernandes, R.; Lubritto, C.; Cocozza, C. IsoMedIta: A Stable Isotope Database for Medieval Italy. *Res. Data J. Humanit. Soc. Sci.* **2023**, *8*, 1–13. [CrossRef]
82. Formichella, G.; Soncin, S.; Lubritto, C.; Tafuri, M.A.; Fernandes, R.; Cocozza, C. Introducing Isotòpia: A Stable Isotope Database for Classical Antiquity. *PLoS ONE* **2024**, *19*, e0293717. [CrossRef]
83. Farese, M.; Soncin, S.; Robb, J.; Fernandes, R.; Tafuri, M.A. The Mediterranean Archive of Isotopic Data, a Dataset to Explore Lifeways from the Neolithic to the Iron Age. *Sci. Data* **2023**, *10*, 917. [CrossRef]
84. di Cicco, M.R.; Altieri, S.; Mantile, N.; Petitti, P.; Persiani, C.; Conti, A.M.; Allegrezza, L.; Cavazzuti, C.; Lubritto, C. Exploring Burial and Dietary Patterns at the Copper Age Necropolis of Selvicciola (Viterbo, Italy): New Perspectives from 14C and Stable Isotope Data. *Heritage* **2024**, *7*, 3291–3309. [CrossRef]

heritage

MDPI

Article

Fission Track Dating of Obsidian Samples from Lipari Neolithic Settlements

Maria Clara Martinelli [1], Letizia Bonizzoni [2,*], Mauro Coltelli [3], Marco Manni [3], Arianna Pefano [2], Massimo Oddone [4] and Alessandra Guglielmetti [2]

1 Parco Archeologico delle Isole Eolie, Museo Luigi Bernabò Brea, Via Castello 2, 98055 Lipari, Italy; mariaclara.martinelli@regione.sicilia.it
2 Department of Physics "Aldo Pontremoli", University of Milan, Via Celoria 16, 20133 Milano, Italy; 906052@stud.unive.it (A.P.); alessandra.guglielmetti@unimi.it (A.G.)
3 Istituto Nazionale di Geofisica e Vulcanologia, Osservatorio Etneo, Piazza Roma 2, 95125 Catania, Italy; mauro.coltelli@ingv.it (M.C.); marco.manni@ingv.it (M.M.)
4 Department of Chemistry, University of Pavia, Via Taramelli 12, 27100 Pavia, Italy; massimo.oddone@unipv.it
* Correspondence: letizia.bonizzoni@unimi.it

Abstract: The present work describes the first results of the project "Lipari Obsidian and Neolithic Human Communities in the Aeolian Islands", which aims to study the connection between obsidian sources on the island of Lipari and Neolithic populations on the Aeolian archipelago in Italy. Obsidian is a natural volcanic glass used to produce chipped tools; in the Neolithic period it was the sharpest known material and its trade played an important role in the Mediterranean area. It is thus of particular interest for tracing prehistoric trading patterns. Indeed, Lipari obsidian has a wide distribution and has been found even in southern France, Dalmatia, Sicily and mainland Italy. To reach the project goal, we considered both raw materials from different obsidian geological samples and artefacts from Neolithic settlements on the Aeolian islands, and performed fission-track dating (FT), a radiometric technique that can be used for uranium-bearing minerals and glasses. The preliminary results facilitated the age determination of geological samples, which we could relate to the different eruption phases. Archaeological samples were also dated; their link with the studied volcanic deposits and lava flows made it possible to shed some new light on raw material procurement and on the ability of the Neolithic populations to move from their locations, with particular attention to the consequences of environmental features on the first human settlements on the Aeolian islands.

Keywords: Neolithic settlements; obsidian flows; fission-track dating

1. Introduction

1.1. The Project

The present work is part of the project "Lipari Obsidian and Neolithic Human Communities in the Aeolian Islands", which concerns the study and analysis of obsidian flows on the island of Lipari in relation to the population of the Aeolian Islands during the Neolithic. The project proposes some scientific aims that concern a synthesis of archaeological and volcanological studies, and at the same time new chronological elements for the temporal reconstruction of the volcanic events that produced obsidian.

The area of interest of the project is the correlation between the first phases of population and the exploitation of this raw material.

The main objectives of the project are:

(1) Analysis of the supply sources (location, dates, sample analysis) of raw material and obsidian artefacts in use in the Aeolian Neolithic settlements, witnessing the volcanic activities of Lipari in prehistory.
(2) Hypotheses on the methods of collecting raw material and the mobility of communities in relation to the Neolithic settlements in the Aeolian Islands.
(3) Environmental characteristics that may have influenced the first population of the islands.

In the present paper we describe how, through the fission-track method, it is possible to determine the age of lava flows and pyroclastic deposits from which the pieces of obsidian used by the Neolithic communities of Lipari were extracted. This will allow us to develop a local chronological framework of the exploitation of the raw material and to compare it with the absolute and cultural chronologies of the prehistory of the Mediterranean.

1.2. The Neolitic of the Aeolian Islands

The archipelago of the Aeolian Islands (Figure 1) includes seven islands, Lipari, Vulcano, Salina, Panarea, Stromboli, Filicudi and Alicudi, and some islets. Except for Vulcano, all the islands and islets have preserved traces of frequentation and settlements starting from the Neolithic [1–5]. The main impetus for the beginning of human settlement in the area was the presence on the island of Lipari of excellent quality obsidian, black, shiny, glassy and perfect for making cutting tools. Control of this raw material allowed for its exchange on a long-distance basis, determining the first form of "trade" in the Mediterranean.

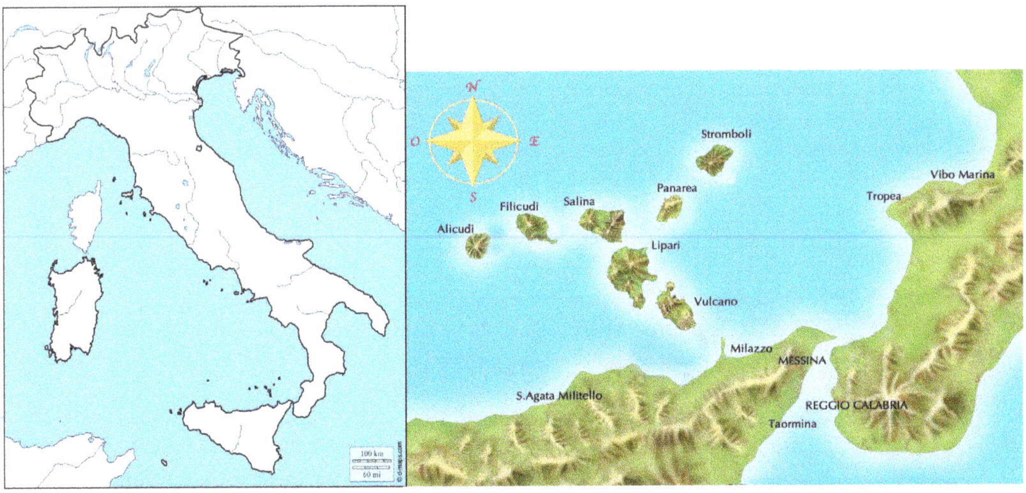

Figure 1. The Aeolian Islands archipelago.

The subaerial eruptive history of Lipari [6] developed between about 267,000 years ago and the Late Middle Ages. There were several eruptions that produced obsidian. The oldest is Monte Guardia (27–24,000 years ago) which seems to have never been used, followed by Canneto Dentro and Vallone del Gabellotto (about 8700–8400 years before present) which were exploited during prehistory. Finally, the formation of the Monte Pilato and Forgia Vecchia volcanoes probably occurred in two phases at the end of the 8th century [6,7] and in the 13th century [8,9].

The actual territory in the north-eastern part of the island is the result of the most recent eruptions and natural erosion that have modified the environment frequented during prehistory. Detailed analyses of the geological sources have allowed us to identify two sources where obsidian was collected in the Neolithic: Vallone del Gabellotto and Canneto Dentro [10], and probably a third source in Vallone Fiume Bianco [11]. Due to the changes in this part of the territory, the areas for collecting raw materials frequented during the Neolithic are no longer visible, but two supply and processing areas have been discovered. The first, Canneto–Lami, relating to the early Neolithic, has been dated through carbon-14 dating at 6290 ± 30 BP (calibrated 5320–5217 BC) in [11]; the second, Canneto–Acquacalda, relating to the late Neolithic, has returned a fragment of red monochrome pottery from the Diana culture [12]. The exploitation of Lipari obsidian began in the Neolithic around the middle of the 6th millennium BC and continued until the Middle Bronze Age (1500–1300 BC), but then disappeared completely. The large quantity of processing chips, blades and exploited cores found in the settlements of the Aeolian Islands, from the Neolithic to the Early Bronze Age, demonstrates that the working of this material took place as a domestic craft within the villages [2]. In the most ancient phase of the population of the Aeolian Islands (Stentinello culture), dated around 5300 BC, two main settlements are known: one in Lipari at the Castellaro site [12] and the other in Salina at the Rinicedda site [13]. Both are characterized by their position in flat areas where subsistence agriculture is possible. Archaeological traces speak of small agricultural communities who lived in huts made of reeds and clay. Their main activity was the exploitation and collection of obsidian. The ceramics that characterize the Stentinello culture are decorated with impressions made by hand or molds. This pottery is associated with the production of fine purified ceramics painted with red or brown banded motifs [14]. The rapid coastal advance of the Neolithic settlers led to the establishment of exchange networks that determined an early circulation of obsidian for hundreds of kilometers, and consequently the Aeolian Neolithic community manifested the need to control the primary source of raw material. In fact, during the periods of painted vases, first trichrome pottery then Serra d'Alto pottery, the settlement stabilized only in Lipari on the coastal cliff, today called a castle, naturally fortified by rocky crags that due to these characteristics have been inhabited continuously over the centuries. During the late Neolithic, the Diana culture emerged, characterized by a red monochrome pottery, widespread in Sicily and the Italian peninsula between 4500 and 4000 BC. Its name derives from the large settlement located on a large coastal plain that was discovered in the middle of the last century in Lipari in the Diana district [15]. During the period of the Diana culture the obsidian chipping technique became highly specialized [16]. All the islands, except Vulcano, were inhabited with settlements; the diffusion of Lipari obsidian reached the entire Mediterranean area and beyond [17].

The end of the Neolithic dates back to the beginning of the 4th millennium BC (Diana–Spatarella culture) and coincided with a period of economic and demographic crisis caused by the introduction of metalworking techniques, which reduced the demand for obsidian, and by the volcanic activities of Stromboli and Vulcano which directly and indirectly influenced the population of Lipari with consequences throughout the archipelago [18].

1.3. *Provenance Studies for Obsidian Samples*

As part of the present project, fission-track (FT) dating of obsidian is utilized as a tool for provenance determination. This approach is widely practiced in the field of archaeometry and has been applied in various contexts, as obsidian holds paramount importance for archaeological studies predating the advent of ceramics [19–22].

Several chemical analyses have demonstrated their effectiveness in distinguishing sources of obsidian since the initial studies conducted in the 1960s [19,23–26]. Early methods included emission spectrometry [27] and proton inelastic scattering [28]. Over time, more advanced techniques were introduced, such as laser ablation–inductively coupled plasma mass spectrometry (LA-ICP-MS) [29], particle-induced X-ray emission (PIXE) [30,31] and energy-dispersive spectroscopy with a scanning electron microscope (SEM-EDS) [32].

One of the most commonly used methods remains neutron activation analysis (NAA) [33], a nuclear technique occasionally paired with particle-induced gamma emission (PIGE) [26] or X-ray fluorescence (XRF) [32]. This unique analytical method offers high precision and detects a wide range of elements; however, it requires irradiating samples with a neutron flux. Over the years, Raman micro-spectroscopy [21,31], Mössbauer spectroscopy [15] and optical emission spectroscopy (OES) [27] have also been employed for similar purposes.

Among the techniques applied to obsidian provenance studies, energy-dispersive X-ray fluorescence (ED-XRF) has emerged as a leading method [16,17,34–38]. It enables the quantification of elements in a non-destructive manner, offering relatively low detection limits for samples of any size. Furthermore, both portable and laboratory-based equipment are available for this purpose. Despite this, the suboptimal detection of low atomic number (Z) elements necessitates specialized calculation and calibration software, requiring customized data processing [39]. This limitation is often addressed by focusing on element ratios rather than absolute elemental concentrations for geographic origin discrimination [16,32,38]. Using this approach, studies on obsidian samples consistently demonstrate the capability of XRF—whether performed with tabletop or portable devices—to reliably identify sample provenance [40,41].

Starting from this premise, it is worth emphasizing the practical advantages of FT dating, particularly its ability to provide absolute age constraints. This capability not only supports provenance determination but also enhances broader archaeological chronologies. While the method necessitates specialized laboratory facilities and a minimum uranium content in the analyzed glass, it serves as a valuable complement to geochemical techniques, especially in complex volcanic settings.

2. Materials and Methods

In 2019, a research agreement was signed between the Archaeological Park of the Aeolian Islands—Luigi Bernabò Brea Museum and the University of Milan—Department of Physics, which implemented some of the aims proposed by the previously mentioned project. The sampling of obsidian flows and tephra layers accessible on the island of Lipari was carried out and, at the same time, samples of obsidian artefacts of the chipping products (debitage) from archaeological excavation contexts in the main Neolithic settlements of the Aeolian archipelago were selected [11].

2.1. Obsidian Samples

Samples from both geological sources and archaeological sites were selected for this study. The geological samples are from dated volcanic phases and comprehend obsidian blocks from pyroclastic deposits and lava flows, or from secondary beds near known or supposed ancient obsidian sources. The archaeological materials, mainly splinters and blades, were from both Salina and Lipari Islands.

2.1.1. Geological Samples

In the present work, 11 geological samples (see Figure 2 for the location of the sampling and Figure 3 for a selection of fragments) have been analyzed; Table 1 summarizes their provenance and eruption age, while further details are given below.

Figure 2. Map of the island of Lipari: the locations where geological samples and Lipari archaeological site samples have been collected are marked in red.

Figure 3. From left to right, obsidian fragments from geological samples 12, 12B and 13A, respectively.

Table 1. Obsidian geological samples used to characterize possible recruitment sites for Neolithic people.

Sample	Sampling Area	Coordinates
LIP2	Canneto Dentro (at a desalination plant) agricultural soil with obsidian in secondary deposit)	38°28′56″ N; 14°57′22″ E
LIP3	Canneto Dentro (thepra layer)	38°29′1″ N; 14°57′37″ E
LIP3A	Canneto Dentro (thepra layer)	38°29′1″ N; 14°57′37″ E
LIP4	Vallone Fiume Bianco (thepra layer)	38°29′44″ N; 14°56′ 2″ E
LIP5	Vallone Fiume Bianco (in the river bed)	38°29′44″ N; 14°56′2″ E
LIP6	Vallone Fiume Bianco (at a modern pumice quarry—thepra layer)	38°29′58″ N; 14°55′49″ E
LIP7	Vallone Gabellotto (Pomiciazzo lava flow)	38°29′42″ N; 14°57′24″ E
LIP12	Monte Guardia (by surface collection)	38°27′42″ N; 14°56′51″ E
LIP12A	Monte Guardia (by surface collection)	38°27′37″ N; 14°56′41″ E
LIP12B	Monte Guardia (by surface collection)	38°27′37″ N; 14°56′41″ E
LIP13A	Vallone Gabellotto (thepra layer from the right slope of the river)	38°29′37″ N; 14°57′29″ E

LIP2, LIP3 and LIP3A are all from the Canneto Dentro area, and have never been directly dated before. LIP2 was extracted in agricultural land within a secondary deposit formed after a landslide. This obsidian sample appears black and shiny despite being immersed in soil material. LIP3 was taken from a pyroclastic deposit formed during the Canneto Dentro dome emplacement and appears black in color with a shiny and compact surface. LIP3A has the same origin as LIP3, and it is characterized by a black color, an opaque appearance and a rough surface, as well as by the presence of bubbles and inclusions. These samples should belong to the same eruptive phase as Vallone del Gabellotto–M. Pilato, Epoch 9a of the geological map [6].

LIP4 and LIP5 are both attributed to the Vallone del Gabellotto Formation. The former is composed of four separate fragments collected from the pumice tephra layer and looks greyish and rough, with many bubbles present in the obsidian itself. LIP5 was collected from the bed of a stream at Vallone Fiume Bianco; the sample has a black color and a shiny surface which presents, in some areas, residues of rocky material. LIP6 was collected from a layer of a pumice quarry in the Vallone Fiume Bianco area, and it is characterized by a black color, a shiny and compact appearance and a smooth surface; inside, it has a large number of bubbles and inclusions. LIP7, from the Pomiciazzo lava flow (on the right bank of the Gabellotto gorge), is black, shiny and relatively opaque, with some inclusion visible under the microscope, and surrounded by a different rocky material. LIP13A is a black and shiny obsidian block, with many inclusions, in a pumice tephra layer from the left bank side of the Gabellotto gorge.

LIP12, LIP12A and LIP12B appear to be opaque, black to gray in color and with a very large number of inclusions. These samples come from the Monte Guardia and Monte Giardina flows and have an expected age of 27–24,000 years BP [6].

2.1.2. Archaeological Samples

Nine archaeological samples (see Table 2 for details and Figure 4 for a selection of fragments) were selected among obsidian artefacts from the Neolithic sites of Lipari and Salina [4], held among Lipari Museum materials, such as splinters and blades of different shapes and sizes, used for various types of daily activities. Details of the samples are given in the following.

Table 2. Obsidian archaeological samples from Neolithic archaeological sites.

Sample	Sampling Area	Excavation Year
MU2	Lipari, Contrada Diana, Trench XVII, Square E, Cut 4	1953
MU4	Lipari Castellaro	1957
MU6	Salina, Rinicedda, Area A, Cuts 2–6	1989
MU7	Salina, Rinicedda, Square L-M 42–44 US 6, Cut 1	2008
MU9	Salina, Rinicedda, Square N43 US 6, Cut 3	2008
MU10	Salina, Rinicedda, Square N44 US 6, Cut 2	2008
MU12	Salina, Rinicedda, Square O44 US 7, Cut 1	2008
MU13	Salina, Rinicedda, Square M45 US 7, Cut 3	2008
MU14	Lipari, Canneto–Acquacalda, neolithic quarry	1977

Figure 4. From left to right, top to bottom: obsidian fragments from archaeological samples MU2, MU4, MU9, MU10, MU12, MU13 and MU14, respectively.

Sample MU2 consists of eight pieces found in 1953 in Lipari in the Contrada Diana district. All samples have an opaque appearance and a black color with gray areas. They present, depending on the fragment under analysis, a variable bubble and inclusion presence, which made fission-track dating a delicate task.

Sample MU4 consists of two blocks excavated in 1957 in Lipari at the Castellaro site. They appear slightly different from each other: one is opaque and with a black color tending towards grey, while the other is blacker and shiny. Under the microscope, the fragments under analysis have bubbles and inclusions inside them.

Samples MU6 (two fragments), MU7 (four fragments), MU9 (three fragments), MU10 (three fragments), MU12 (five fragments) and MU13 (two fragments) are from the Rinicedda archaeological site on Salina Island excavated in 1989 and 2008. They are all opaque and black, with bubbles and inclusions. Sample MU7 shows the presence, in some areas, of rocky material; MU13 is grayish.

Sample MU14 (three fragments) was recovered from a Neolithic quarry on the Canneto–Acquacalda road. The fragments are black and characterized internally by the presence of bubbles and inclusions.

2.2. *Fission-Track Dating*

Obsidian, a naturally occurring volcanic glass formed when lava extruded from a volcano cools rapidly with minimal crystal growth, can be dated by means of the fission-track method, which has already proved to be useful for source identification [42,43] and provenance studies [44,45]. In Neolithic times, obsidian was the sharpest known material, and it is thus considered one of the earliest pieces of evidence for commerce conducted over a long distance.

Fission-track dating (FTD) [46–48] is based on the detection of the fission products of uranium that create damage trails (fission tracks) within minerals [49,50]. It is a radiometric technique used for both very old materials, such as meteorites, and younger objects, including artifacts from archaeological sites, providing a versatile tool for dating a wide range of samples. In fission-track dating, the fission products are identified as microscopic damage trails or tracks in the crystal lattice rather than as daughter isotopes. ^{238}U undergoes spontaneous fission decay, and it is the only isotope with a decay rate relevant to the significant production of natural fission tracks. As detailed in Section 2.1.1, the density of spontaneous fission tracks in a material increases over time as events accumulate, making it directly proportional to the age of the sample. In general, obsidian is datable by FTD [49,51–53], and thus this technique is a tool for reconstructing the chronological framework of Lipari obsidian flows. As collecting and processing obsidian was the main economic resource for the Neolithic population of Lipari, FTD on archaeological findings allows us to correlate obsidian artifacts with natural sources.

The physical bases of FTD rely on the fact that the spontaneous nuclear fission of uranium 238 within minerals creates damage trails, the so-called fission tracks, that can be etched and observed with an optical microscope [54–56]: the surface density of fission tracks on a thin chip of material thus represents a measure of the time over which tracks accumulate. Both spontaneous and induced fission tracks are uniformly distributed throughout the entire volume of the material under study, as the uranium that generates them is uniformly distributed. However, the counting is performed on an internal surface obtained by sectioning the material. For this reason, we refer to surface density.

To calculate the fission-track age, it is necessary to measure the amount of uranium present by irradiating the sample with thermal neutrons in a nuclear reactor. This procedure artificially induces fission on uranium 235; the ratio between uranium 238 and uranium 235 fission-track density is used to calculate the age.

2.2.1. Fission-Track Age Equation

The fission-track method is based on the natural radioactive decay of the unstable parent nucleus, uranium 238, into a stable daughter nucleus, the fission fragments. The age equation in its most complete form is given by:

$$T = \frac{1}{\lambda_\alpha} ln\left(1 + \frac{\lambda_\alpha}{\lambda_f}\frac{\sigma\phi}{\eta}\frac{N_s}{N_i}\right), \quad (1)$$

where:

λ_α is the decay constant of uranium 238 for alpha decay;

N_s/N_i is the ratio between the number of uranium 238 nuclei per unit volume that decay by means of spontaneous fission and the number of uranium 235 nuclei per unit volume that decay by means of induced fission after neutron irradiation. The ratio can be determined through the ratio of surface densities of fission tracks determined in the non-irradiated and irradiated samples.

For samples younger than 100 million years (Ma), that is, for the vast majority of the studied materials, the approximate form of the age equation is used:

$$T = \frac{\phi\sigma}{\lambda_f\eta}\cdot\frac{\rho_s}{\rho_i}, \quad (2)$$

where:

T is the age in years,

ρ_s is the surface density of fossil tracks,

and ρ_i is the surface density of induced tracks.

From Equation (2), it can be clearly seen that T is proportional to the surface density of fossil and induced tracks, and thus to the uranium content in the sample, as well as to the thermal neutron dose and a series of constants, namely [57]:

- Φ is the thermal neutron dose (neutrons/cm^2), which can take different values depending on the dosimeter used and the characteristics of the nuclear reactor;
- σ is the neutron cross-section for the fission of uranium 235; this is not a true constant but is an average associated with the spectrum of thermal neutrons produced in nuclear reactors under standard conditions;
- λ_f is the decay constant for the spontaneous fission of uranium 238; this value includes some uncertainty due to the low probability of the event;
- η is the isotopic ratio of uranium 238 to uranium 235.

The values of the above-described parameters used in the present research are reported in Table 3.

Table 3. Numerical values of the parameters to be used in the age equation.

Φ	1.48×10^{15} n/cm^2
σ	580.2 barn
λ_f	7.03×10^{-17} y^{-1}
η	137.88

The age Equation (2) is valid if several important assumptions are met [58], i.e., if it is assumed that every single fission event leaves a latent track in the material and that this track remains stable over time; that the system is considered "closed", meaning there have been no variations in the uranium content of the sample; if the uranium content is assumed to be uniformly distributed throughout the material; and if the fission tracks from the spontaneous fission of uranium 238 and the induced fission tracks of uranium 235 are considered equivalent, as they have a negligible difference in energy.

The total error value in the age equation is given by the sum of the errors encountered in the calculation of the individual variables, namely those related to the dose Φ and the ratio ρ_s/ρ_i, assuming λ, σ and η as constants.

2.2.2. Experimental Procedure

Samples for FTD require long and laborious preparation, consisting of a series of steps that can be summarized as follows. Firstly, the obsidian block is broken into thin chips, divided into two batches, one of which is irradiated with thermal neutrons. The neutron irradiation took place using the TRIGA MARK II research reactor at the L.E.N.A. laboratory at the University of Pavia. The reactor is fueled with low-enriched 235U and moderated with light demineralized water, with a steady-state power of 250 kW. For the irradiation facility, the rotating sample holder, LS (Lazy Susan), was selected in order to achieve a thermal flux around 1×10^{12} n cm^{-2} s^{-1} [59].

After irradiating part of the samples, the chips from both batches are embedded into epoxy resin, so as to obtain resin cylinders that serve as support for the processing. For this purpose, a two-component epoxy resin (Buehler), which polymerizes at room temperature, was used. Grinding and polishing are the last two stages of sample preparation before etching; the chemical treatment, employed to selectively etch the obsidian surface, enlarges the fission tracks and allows a clearer detection through optical microscope observation.

The standard conditions for the chemical treatment of obsidian are HF in a 20% aqueous solution at a temperature of 40 °C for 120 s.

At the end of this procedure, it is possible to count the fossil and induced traces and to perform the calculations for age determinations.

3. Results

The results obtained are summarized in Figure 5, which shows the comparison between the dates obtained for the geological samples and the archaeological samples. More details and the discussion of the results are provided in Sections 3.1 and 3.2.

Table 4. FTD ages obtained for obsidian geological samples used to characterize possible recruitment sites for Neolithic people.

Sample	Sampling Area	Stratigraphy Age (y)	FTD Age (y)
LIP2	Canneto Dentro	9000–8000 *	9714 ± 5610
LIP3	Canneto Dentro	9000–8000	10,587 ± 3747
LIP3A	Canneto Dentro	9000–8000	9265 ± 6557
LIP4	Vallone Fiume Bianco	9000–8000	9491 ± 1885
LIP5	Vallone Fiume Bianco	9000–8000	10,473 ± 6049
LIP6	Vallone Fiume Bianco	9000–8000	10,019 ± 4101
LIP7	Vallone Gabellotto	8700–8400	10,183 ± 2739
LIP12	Monte Guardia	27,000–24,000	21,306 ± 4778
LIP12A	Monte Guardia	27,000–24,000	25,117 ± 5977
LIP12B	Monte Guardia	27,000–24,000	28,969 ± 8086
LIP13A	Vallone Gabellotto	8700–8400	11,922 ± 3208

* The age of Canneto Dentro has been assumed from geological studies, but never directly dated until the present work.

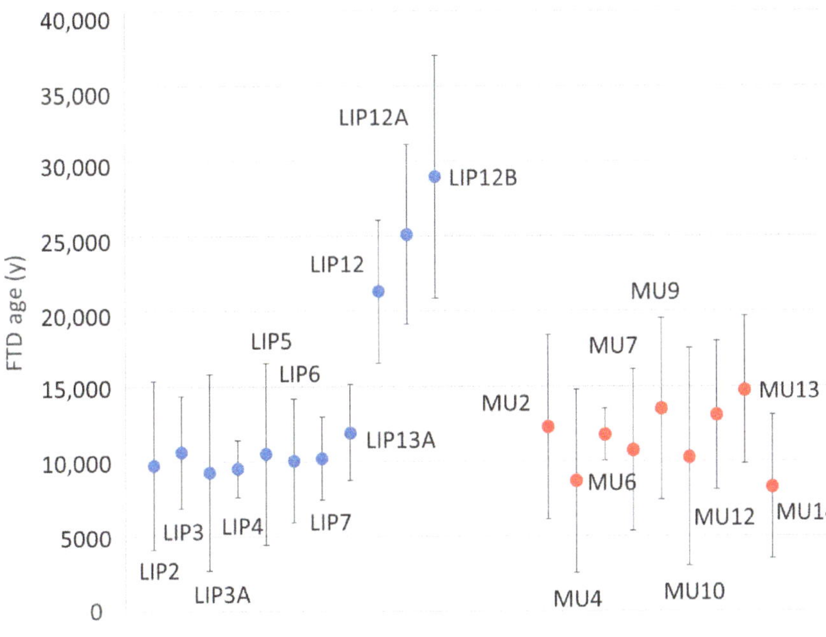

Figure 5. Graphical summary of the results outlined in Tables 3 and 4, highlighting the relationship between the ages of obsidian geological samples (blue dots) and archaeological findings (red dots).

3.1. Geological Samples

The results obtained for the geological samples analyzed in this work are reported in Table 4.

The most ancient geological samples (LIP12, LIP12A and LIP12B) are in agreement with the volcanism activity in southern Lipari—Epoch 8a dated 27–24,000 years before present [6]. The other samples belong to the Vallone del Gabellotto Formation, whose stratigraphic interval age is still uncertain. The Canneto Dentro tephra, associated with the obsidian dome, is placed in Epoch 8b, recognized as occurring between the older layers of the Upper Brown Tuffs (24–20 ka) and the Vallone del Gabellotto (8.7–8.4 ka) [6,60]. In recent literature, the date 8.6 ka is assumed as the most probable age of the Vallone del Gabellotto Formation that included the Pomiciazzo obsidian lava flow; however, another date was obtained with the fission-track method, 11,400 ± 1800 BP [44], on obsidian samples from Vallone del Gabellotto. The samples LIP4, LIP5 and LIP6 belong to a long period of volcanic activity but concern an area distant from the primary source of the Vallone del Gabellotto, located where the Fiume Bianco torrent digs the tephra layers of pumice belonging to the Vallone del Gabellotto Formation, uncovering the obsidian blocks present in that deposit dating up to 7170 years BP [61].

3.2. Archaeological Samples

Table 5 shows the fission-track datings obtained for the archaeological samples that are clearly related to the obsidian taken from the Vallone del Gabellotto formation. In the most ancient settlements, Castellaro (Lipari) and Rinicedda (Salina), the obsidian was probably collected from two main sources: (1) the bed of the near river Fiume Bianco and (2) the first Vallone del Gabellotto quarry [11]. In the more recent Neolithic settlement, Contrada Diana, the obsidian was collected primarily in the Vallone del Gabellotto area that included the Canneto Dentro source [10,17]. The most ancient dates are reported from samples of the Rinicedda site (MU6, MU12, MU13). We could suppose that this obsidian belongs to a more ancient eruption which is no longer visible. It is possible that the Neolithic community first inhabited Salina Island and then Lipari Island. This hypothesis could agree with the observations made by Buchner in 1949 [62] in the Papesca locality (actually Capo Rosso) where he observed the presence of older lava at the base of the stratigraphy [63].

Table 5. FTD ages obtained for obsidian archaeological samples from the Lipari and Salina sites.

Sample	Archaeological Site	Excavation Context	FTD Age (y)
MU2	Lipari, Contrada Diana	Trench XVII, E (1953)	12,276 ± 6148
MU4	Lipari Castellaro	(1957)	8660 ± 6129
MU6	Salina, Rinicedda	Square A (1989)	11,749 ± 1707
MU7	Salina, Rinicedda	Square L-M 42–44 US 6 (2008)	10,726 ± 5371
MU9	Salina, Rinicedda	Square N43 US6 (2008)	13,461 ± 6031
MU10	Salina, Rinicedda	Square N44 US6 (2008)	10,261 ± 7261
MU12	Salina, Rinicedda	Square O44 US7 (2008)	13,063 ± 4951
MU13	Salina, Rinicedda	Square M45 US7 (2008)	14,740 ± 4935
MU14	Lipari, Canneto–Acquacalda	Second quarry (1977)	8256 ± 4772

The MU14 sample was collected in the Neolithic second quarry, discovered by Buchner near the Canneto–Acquacalda road, used during the Diana culture phase. This is the youngest sample among the archaeological ones dated in this work. The fission-track date accorded with the chronology of the Vallone del Gabellotto formation but the sample MU2 from Contrada Diana has a different and older date.

Considering that the C14 dating carried out in the archaeological sites indicates the beginning of the population of the Aeolian Islands in the middle of the VI millennium BC, it is assumed that the period of quiescence of the volcanoes of Lipari had begun. In the Middle Ages, the volcanic activity resumed, modifying the territory of the north-eastern part of the island.

Comparing the dates of geological and archaeological samples, a higher incidence is observed around 10,000 years ago, suggesting the presence of an eruptive phase of the Vallone del Gabellotto formation older than the one dated at 8700–8400 years but not yet identified.

4. Conclusions

After the first outcome of this project, which focuses on the dating of geological and archaeological samples using the fission-track method and is the first of its kind to be proposed, we can summarize our conclusive remarks as follows:

These findings may support the hypothesis that obsidian from Lipari was collected and used in early Neolithic (impressed ware) or even Mesolithic settlements outside Lipari, even before the stable occupation of the Aeolian Islands.

The preliminary results presented in the present paper clearly show the potential of the fission-track dating method for provenance studies, particularly in cases where geochemically similar obsidian sources cannot be distinguished using traditional methods. Moreover, its ability to provide absolute age constraints contributes not only to provenance determination but also to broader archaeological chronologies.

The project "Lipari Obsidian and Neolithic Human Communities in the Aeolian Islands" is currently ongoing, aiming to further investigate and deepen our understanding of these aspects.

Author Contributions: Conceptualization, M.C.M., M.C. and A.G.; methodology, M.C.M., M.C., M.M., M.O. and A.G.; investigation, L.B., A.P. and A.G.; data curation, L.B., A.P. and A.G.; writing—original draft preparation, M.C.M., M.M., L.B. and A.P.; writing—review and editing, M.C.M., M.C., M.M., L.B. and A.G.; supervision, M.C.M., M.C. and A.G. All authors have read and agreed to the published version of the manuscript.

Funding: This research received no external funding.

Data Availability Statement: The raw data supporting the conclusions of this article will be made available by the authors on request.

Conflicts of Interest: The authors declare no conflicts of interest.

References

1. Forni, F.; Lucchi, F.; Peccerillo, A.; Tranne, C.A.; Rossi, P.L.; Frezzotti, M.L. Chapter 10 Stratigraphy and Geological Evolution of the Lipari Volcanic Complex (Central Aeolian Archipelago). *Geol. Soc. Lond. Mem.* **2013**, *37*, 213–279. [CrossRef]
2. Martinelli, M.C.; Dawson, H.; Cascio, P.L.; Levi, S.T.; Fiorentino, G. Blowin' in the Wind: Settlement, Landscape and Network Dynamics in the Prehistory of the Aeolian Islands. *J. Mediterr. Archaeol.* **2021**, *34*, 28–57. [CrossRef]
3. Martinelli, M.C. *Isole Vicine. L'arcipelago delle Isole Eolie e le Comunità Umane Nella Preistoria Mediterranea*; Ediz. Storia e Studi Sociali: Ragusa, Italy, 2020; ISBN 978-88-99168-50-6.
4. Bernabò Brea, L.; Cavalier, M. Civiltà Preistoriche Delle Isole Eolie e Del Territorio Di Milazzo. *Bull. Paletnol.* **1956**, *65*, 8–98.
5. Martinelli, M.C.; Lo Cascio, P. Topografia Della Preistoria Nelle Isole Eolie. In *A Madeleine Cavalier*; Bernabò Brea, M., Massimo Cultraro, M., Gras, M., Martinelli, M.C., Pouzaudoux, C., Spigo, U., Eds.; Collection du Centre Jean Berard 49, CNRS: Napoli, Italy, 2018; pp. 65–78.
6. Di Renzoni, A.; Levi, S.; Renzulli, A.; Rosi, M.; Yoon, D. Should I Stay or Should I Go? 6000 Years of Human Presence and Abandonments at Stromboli Volcano and an Overview on the Whole Aeolian Archipelago (Southern Tyrrhenian Sea, Italy). *Ann. Geophys.* **2021**, *64*, VO545. [CrossRef]
7. Keller, J. Lipari's Fiery Past: Dating the Medieval Pumice Eruption of Monte Pelato. In Proceedings of the International Conference 'The Fire Between Air and Water', Lipari, Italy, 23 September–2 October 2002.

8. Tanguy, J.-C.; Le Goff, M.; Principe, C.; Arrighi, S.; Chillemi, V.; Paiotti, A.; La Delfa, S.; Patanè, G. Archeomagnetic Dating of Mediterranean Volcanics of the Last 2100 Years: Validity and Limits. *Earth Planet. Sci. Lett.* **2003**, *211*, 111–124. [CrossRef]
9. Pistolesi, M.; Rosi, M.; Malaguti, A.B.; Lucchi, F.; Tranne, C.A.; Speranza, F.; Albert, P.G.; Smith, V.C.; Di Roberto, A.; Billotta, E. Chrono-Stratigraphy of the Youngest (Last 1500 Years) Rhyolitic Eruptions of Lipari (Aeolian Islands, Southern Italy) and Implications for Distal Tephra Correlations. *J. Volcanol. Geotherm. Res.* **2021**, *420*, 107397. [CrossRef]
10. Tykot, R.H.; Feund, K.P.; Vianello, A. Source Analysis of Prehistoric Obsidian Artifacts in Sicily (Italy) Using pXRF. In *Archaeological Chemistry VIII*; ACS Symposium Series; Armitage, R.A., Burton, J.H., Eds.; American Chemical Society: Washington, DC, USA, 2013.
11. Martinelli, M.C.; Coltelli, M.; Manni, M.; Bonizzoni, L.; Guglielmetti, A.; Oddone, M.; Balestrieri, M.L. Prehistorical Obsidian Sources in the Island of Lipari (Aeolian Islands). *Open Archaeol.* **2020**, *6*, 393–402. [CrossRef]
12. Cavalier, M. Ricerche Preistoriche Nell'arcipelago Eoliano. *Riv. Sci. Preist. Ist. Ital. Preist. Protostoria Firenze* **1979**, *XXXIV*, 45–135.
13. Bernabò Brea, L.; Cavalier, M. Salina. Ricerche Archeologiche (1989–1993). In *Meligunis Lipara VIII*; Accademia di Scienze Lettere e Arti, Regione Siciliana; Assessorato dei Beni Culturali: Palermo, Italy, 1995.
14. Armetta, F.; Giuffrida, D.; Martinelli, M.C.; Mollica Nardo, V.; Saladino, M.L.; Ponterio, R.C. Non-Invasive Investigation on Pigments of the Aeolian Islands Neolithic Pottery. *Mater. Lett.* **2023**, *336*, 133854. [CrossRef]
15. Cavalier, M.; Brea, L.B.; Luigi, B.B.; Meligunís Lipàra. Vol. 1: La Stazione Preistorica Della Contrada Diana e La Necropoli Protostorica Di Lipari—LibroCo. Available online: https://www.libroco.it/dl/Madeleine-Cavalier-Luigi-BernabA-Brea-Bernabo-Brea-Luigi/Dario-Flaccovio/isbn/Meligunis-Lipara-Vol-1-La-stazione-preistorica-della-contrada-Diana-e-la-necropoli-protostorica-di-Lipari/cw594081988598270.html (accessed on 9 January 2025).
16. Martinelli, M.C.; Tykot, R.H.; Vianello, A. Lipari (Aeolian Islands) Obsidian in the Late Neolithic. Artifacts, Supply and Function. *Open Archaeol.* **2019**, *5*, 46–64. [CrossRef]
17. Tykot, R.H. Geological Sources of Obsidian on Lipari and Artifact Production and Distribution in the Neolithic and Bronze Age Central Mediterranean. *Open Archaeol.* **2019**, *5*, 83–105. [CrossRef]
18. Manni, M.; Coltelli, M.; Martinelli, M.C. Volcanic Events That Have Marked the Anthropic History of the Aeolian Islands. *Ann. Geophys.* **2019**, *62*, VO08. [CrossRef]
19. Torrence, R.; Neall, V.; Doelman, T.; Rhodes, E.; McKee, C.; Davies, H.; Bonetti, R.; Guglielmetti, A.; Manzoni, A.; Oddone, M.; et al. Pleistocene Colonisation of the Bismarck Archipelago: New Evidence from West New Britain. *Archaeol. Ocean.* **2004**, *39*, 101–130. [CrossRef]
20. Chataigner, C.; Badalian, R.; Bigazzi, G.; Cauvin, M.-C.; Jrbashian, R.; Karapetian, S.G.; Norelli, P.; Oddone, M.; Poidevin, J.-L. Provenance Studies of Obsidian Artefacts from Armenian Archaeological Sites Using the Fission-Track Dating Method. *J. Non-Cryst. Solids* **2003**, *323*, 167–171. [CrossRef]
21. Bellot-Gurlet, L.; Dorighel, O.; Poupeau, G. Obsidian Provenance Studies in Colombia and Ecuador: Obsidian Sources Revisited. *J. Archaeol. Sci.* **2008**, *35*, 272–289. [CrossRef]
22. Bigazzi, G.; Ercan, T.; Oddone, M.; Özdoğan, M.; Yeğingil, Z. Application of Fission Track Dating to Archaeometry: Provenance Studies of Prehistoric Obsidian Artifacts. *Nucl. Tracks Radiat. Meas.* **1993**, *22*, 757–762. [CrossRef]
23. Summerhayes, G. Obsidian Network Patterns in Melanesia—Sources, Characterization and Distribution. *Bull. Indo-Pac. Prehist. Assoc.* **2009**, *29*, 109–123. [CrossRef]
24. Stewart, S.J.; Cernicchiaro, G.; Scorzelli, R.B.; Poupeau, G.; Acquafredda, P.; De Francesco, A. Magnetic Properties and 57Fe Mössbauer Spectroscopy of Mediterranean Prehistoric Obsidians for Provenance Studies. *J. Non-Cryst. Solids* **2003**, *323*, 188–192. [CrossRef]
25. Golitko, M.; Meierhoff, J.; Terrell, J.E. Chemical Characterization of Sources of Obsidian from the Sepik Coast (PNG). *Archaeol. Ocean.* **2010**, *45*, 120–129. [CrossRef]
26. Gratuze, B.; Barrandon, J.N.; Isa, K.A.; Cauvin, M.C. Non-Destructive Analysis of Obsidian Artefacts Using Nuclear Techniques: Investigation of Provenance of Near Eastern Artefacts. *Archaeometry* **1993**, *35*, 11–21. [CrossRef]
27. Cann, J.R.; Renfrew, C. The Characterization of Obsidian and Its Application to the Mediterranean Region. *Proc. Prehist. Soc.* **1964**, *30*, 111–133. [CrossRef]
28. Coote, G.E.; Whitehead, N.E.; McCallum, G.J. A Rapid Method of Obsidian Characterisation by Inelastic Scattering of Protons. *J. Radioanal. Chem.* **1972**, *12*, 491–496. [CrossRef]
29. Barca, D.; De Francesco, A.M.; Crisci, G.M. Application of Laser Ablation ICP-MS for Characterization of Obsidian Fragments from Peri-Tyrrhenian Area. *J. Cult. Herit.* **2007**, *8*, 141–150. [CrossRef]
30. Acquafredda, P.; Andriani, T.; Lorenzoni, S.; Zanettin, E. Chemical Characterization of Obsidians from Different Mediterranean Sources by Non-Destructive SEM-EDS Analytical Method. *J. Archaeol. Sci.* **1999**, *26*, 315–325. [CrossRef]
31. Bellot-Gurlet, L.; Bourdonnec, F.-X.L.; Poupeau, G.; Dubernet, S. Raman Micro-Spectroscopy of Western Mediterranean Obsidian Glass: One Step towards Provenance Studies? *J. Raman Spectrosc.* **2004**, *35*, 671–677. [CrossRef]

32. Chemical Fingerprinting and Source Tracing of Obsidian: The Central Mediterranean Trade in Black Gold | Accounts of Chemical Research. Available online: https://pubs.acs.org/doi/10.1021/ar000208p (accessed on 5 February 2025).
33. Knight, C.L.F.; Hu, H.; Glascock, M.D.; Nelson, S.A. Obsidian Sub-Sources at the Zaragoza-Oyameles Quarry in Puebla, Mexico: Similarities with Altotonga and Their Distribution Throughout Mesoamerica. *Lat. Am. Antiq.* **2017**, *28*, 46–65. [CrossRef]
34. Acquafredda, P.; Micheletti, F.; Muntoni, I.M.; Pallara, M.; Tykot, R.H. Petroarchaeometric Data on Antiparos Obsidian (Greece) for Provenance Study by SEM-EDS and XRF. *Open Archaeol.* **2019**, *5*, 18–30. [CrossRef]
35. Moutsiou, T. A Compositional Study (pXRF) of Early Holocene Obsidian Assemblages from Cyprus, Eastern Mediterranean. *Open Archaeol.* **2019**, *5*, 155–166. [CrossRef]
36. Tykot, R.H. A Decade of Portable (Hand-Held) X-Ray Fluorescence Spectrometer Analysis of Obsidian in the Mediterranean: Many Advantages and Few Limitations. *MRS Adv.* **2017**, *2*, 1769–1784. [CrossRef]
37. Characterization of the Monte Arci (Sardinia) Obsidian Sources—ScienceDirect. Available online: https://www.sciencedirect.com/science/article/pii/S0305440396901302?via=ihub (accessed on 5 February 2025).
38. Freund, K.P.; Tykot, R.H.; Vianello, A. Blade production and the consumption of obsidian in Stentinello period Neolithic Sicily. *Comptes Rendus Palevol* **2015**, *14*, 207–217. [CrossRef]
39. High-Precision Non-Destructive X-Ray Fluorescence Method Applicable to Establishing the Provenance of Obsidian Artifacts—Giauque—1993—X-Ray Spectrometry—Wiley Online Library. Available online: https://analyticalsciencejournals.onlinelibrary.wiley.com/doi/10.1002/xrs.1300220111 (accessed on 5 February 2025).
40. Bonizzoni, L.; Kulchytska, O.; Ruschioni, G. XRF Semi-Quantitative Analysis and Multivariate Statistics for the Classification of Obsidian Flows in the Mediterranean Area. *Appl. Sci.* **2023**, *13*, 3495. [CrossRef]
41. Tykot, R.H. Non-Destructive pXRF on Prehistoric Obsidian Artifacts from the Central Mediterranean. *Appl. Sci.* **2021**, *11*, 7459. [CrossRef]
42. Bigazzi, G.; Màrton, P.; Norelli, P.; Rozložnik, L. Fission Track Dating of Carpathian Obsidians and Provenance Identification. *Int. J. Radiat. Appl. Instrum. Part D Nucl. Tracks Radiat. Meas.* **1990**, *17*, 391–396. [CrossRef]
43. Bigazzi, G.; Hadler Neto, J.C.; Iunes, P.J.; Osório Araya, A.M. Fission-Track Dating of South American Natural Glasses: An Overview. *Radiat. Meas.* **2005**, *39*, 585–594. [CrossRef]
44. Bigazzi, G.; Bonadonna, F. Fission Track Dating of the Obsidian of Lipari Island (Italy). *Nature* **1973**, *242*, 322–323. [CrossRef]
45. Ambrose, S.H. Obsidian Dating and Source Exploitation Studies in Africa: Implications for the Evolution of Human Behavior. In *Obsidian and Ancient Manufactured Glasses*; University of New Mexico Press: Albuquerque, NM, USA, 2012; pp. 56–72.
46. Westgate, J.; Sandhu, A.; Shane, P. Fission-Track Dating. In *Chronometric Dating in Archaeology*; Taylor, R.E., Aitken, M.J., Eds.; Springer: Boston, MA, USA, 1997; pp. 127–158. ISBN 978-1-4757-9694-0.
47. Wagner, G.A. Fission Track Dating of Apatites. *Earth Planet. Sci. Lett.* **1968**, *4*, 411–415. [CrossRef]
48. Gleadow, A.J.W. Fission-Track Dating Methods: What Are the Real Alternatives? *Nucl. Tracks* **1981**, *5*, 3–14. [CrossRef]
49. Westgate, J.; Naeser, N.; Alloway, B. Fission-Track Dating. In *Encyclopedia of Quaternary Science*; Springer: Dordrecht, The Netherlands, 2013; pp. 643–662.
50. Fitzgerald, P.G.; Gleadow, A.J.W. Fission-Track Geochronology, Tectonics and Structure of the Transantarctic Mountains in Northern Victoria Land, Antarctica. *Chem. Geol. Isot. Geosci. Sect.* **1988**, *73*, 169–198. [CrossRef]
51. Liritzis, I.; Laskaris, N. Fifty Years of Obsidian Hydration Dating in Archaeology. *J. Non-Cryst. Solids* **2011**, *357*, 2011–2023. [CrossRef]
52. Liritzis, I.; Stevenson, C.M. *Obsidian and Ancient Manufactured Glasses*; UNM Press: Albuquerque, NM, USA, 2012; ISBN 978-0-8263-5161-6.
53. Clark, D.L. The Obsidian Dating Method. *Curr. Anthropol.* **1961**, *2*, 111–114. [CrossRef]
54. Price, P.B.; Walker, R.M. Fossil Tracks of Charged Particles in Mica and the Age of Minerals. *J. Geophys. Res.* **1963**, *68*, 4847–4862. [CrossRef]
55. Fleischer, R.L.; Price, P.B. Techniques for Geological Dating of Minerals by Chemical Etching of Fission Fragment Tracks. *Geochim. Cosmochim. Acta* **1964**, *28*, 1705–1714. [CrossRef]
56. Price, P.B.; Walker, R.M. Chemical Etching of Charged-Particle Tracks in Solids. *J. Appl. Phys.* **1962**, *33*, 3407–3412. [CrossRef]
57. Hurford, A.J.; Green, P.F. A Users' Guide to Fission Track Dating Calibration. *Earth Planet. Sci. Lett.* **1982**, *59*, 343–354. [CrossRef]
58. Tagami, T.; O'Sullivan, P.B. Fundamentals of Fission-Track Thermochronology. *Rev. Mineral. Geochem.* **2005**, *58*, 19–47. [CrossRef]
59. Di Luzio, M.; Oddone, M.; Prata, M.; Alloni, D.; D'Agostino, G. Measurement of the Neutron Flux Parameters f and α at the Pavia TRIGA Mark II Reactor. *J. Radioanal. Nucl. Chem.* **2017**, *312*, 75–80. [CrossRef]
60. Zanchetta, G.; Sulpizio, R.; Roberts, N.; Cioni, R.; Eastwood, W.J.; Siani, G.; Caron, B.; Paterne, M.; Santacroce, R. Tephrostratigraphy, Chronology and Climatic Events of the Mediterranean Basin during the Holocene: An Overview. *Holocene* **2011**, *21*, 33–52. [CrossRef]
61. Bigazzi, G.; Coltelli, M.; Norelli, P. Nuove Età Delle Ossidiane Di Lipari Determinate Con Il Metodo Delle Tracce Di Fissione. In Proceedings of the GeoItalia, Bellaria-Igea Marina, Italy, 18 September 2003.

62. Buchner, G. Ricerche Sui Giacimenti e Sulle Industrie Di Ossidiana in Italia. *Riv. Sci. Preist.* **1949**, *IV*, 162–186.
63. Martinelli, M.C.; Manni, M. The Prehistoric Obsidian of Lipari: Gabellotto and the so-Called Pomiciazzo Lava Flow (Lipari, Aeolian Islands). In Proceedings of the 6th Conference A. Rittmann, Catania, Italy, 18–20 September 2024.

heritage

MDPI

Article

A Semi-Automated Machine-Learning Tool for Assessing Building Phases: Discriminant Analysis of Mortars from the 2022 Excavation at the Sarno Bath Complex in Pompeii

Simone Dilaria [1,2,*], Caterina Previato [1,*], Michele Secco [1,2] and Maria Stella Busana [1]

[1] Department of Cultural Heritage: Archaeology, History of Art, Cinema and Music (DBC), University of Padua, Piazza Capitaniato 7, 35139 Padua, Italy; michele.secco@unipd.it (M.S.); mariastella.busana@unipd.it (M.S.B.)
[2] Inter-Departmental Research Centre for the Study of Cement Materials and Hydraulic Binders (CIRCe), University of Padua, Via Giovanni Gradenigo 6, 35131 Padua, Italy
* Correspondence: simone.dilaria@unipd.it (S.D.); caterina.previato@unipd.it (C.P.)

Abstract: This study presents the results of the analyses of 15 structural mortars from the building at civ. 21, level +0 of the Sarno Bath complex in Pompeii. These samples were collected during recent stratigraphic excavations (year 2022) for detailed in-laboratory compositional characterization, aiming to trace the construction phases of the originating walls. The 2022 samples were firstly analyzed via quantitative phase analysis–X-ray powder diffraction. The resulting quantitative mineralogical profiles were then processed alongside those analyzed in previous studies from level +0 structures of the Sarno Baths using multivariate statistical methods, including principal component analysis (PCA) and discriminant analysis, applied to quantitative phase analysis (QPA)–X-ray powder diffraction data (XRPD), to identify and map the construction phases. This approach enabled the correlation of the 2022 samples with previously established construction phases. Polarized-light optical microscopy and scanning electron microscopy (SEM) coupled with energy dispersive X-ray spectroscopy (EDS) were then primarily used for validation purposes. These methods highlighted the compositional differences between samples and revealed significant features related to the use of specific raw materials. These results confirm the reliability of the semi-automated sample processing proposed in this research, adopting discriminant analysis as a machine-learning-based tool for defining construction phases in Pompeian contexts.

Keywords: ancient mortars; QPA-XRPD; discriminant analysis; principal component analysis; Somma–Vesuvius; Phlegraean Fields

1. Introduction

Archaeometric research on ancient mortars, investigated by adopting a diverse array of analytical techniques, has become increasingly common in the last decade. The selection of analytical methods is guided by the research objectives and the study's context. At Pompeii, mortar analyses have often been employed to examine compositional characteristics from a diachronic perspective or in relation to their functional application. A central aim in the compositional analysis of Vesuvian mortars has been to emphasize petrographic and mineralogical variations, facilitating the identification and chronological sequencing of construction phases in specific buildings. These tasks were usually achieved by adopting traditional approaches, primarily based on polarized-light optical microscopy investigations [1–4]. Other studies have combined multiple analytical techniques to distinguish mortars by their production chronology and corresponding construction phases, leveraging specific petrographic, mineralogical, and geochemical variations [5–8]. Recently, the complex building history of the Sarno Bath complex, studied by researchers from the University of Padua, has offered a unique opportunity to investigate a lesser-known building at Pompeii. The construction phases of this building had remained uncertain due to the absence of detailed modern excavation reports and archival documentation from early-20th-century restorations. Therefore, mortar analysis has been used to distinguish the construction phases of the building by examining their compositional characteristics. Secco et al. [9] and Dilaria et al. [10] centered their analyses on a mineralogical characterization approach utilizing QPA-XRPD, which identified consistent mineralogical assemblages indicative of building phase sequences. These mineralogical assemblages were properly framed by processing via PCA (principal component analysis) of the QPA-XRPD data, which successfully differentiated mortar groups according to reliable ancient and modern construction and restoration phases. By characterizing mortar groups aligned with the building phases of the complex, this approach offers valuable insights into the construction history of the Sarno Baths.

The current research builds upon previous studies and aims to analyze a new dataset of mortar samples collected during stratigraphic excavations at the house located at civ. 21 (level +0) of the Sarno Bath complex (*Regio* VIII, *Insula* 2, n. 17–21) (Figure 1a,b).

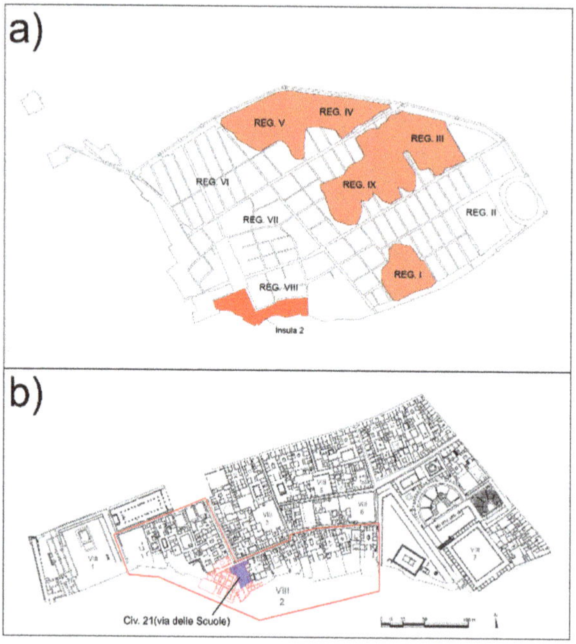

Figure 1. (**a**) General plan of Pompeii with indication of *Insula* 2 in *Regio* VIII; (**b**) Indication of the house at civ. 21 in Via delle Scuole, within the Sarno Bath Complex (modified from [10]).

The recent excavations, conducted in 2022 by the Department of Cultural Heritage at the University of Padua, were carried out in collaboration with the Archaeological Park of Pompeii, under the scientific coordination of M.S. Busana. The samples were initially analyzed using QPA-XRPD, and their mineralogical quantitative profiles were integrated into the existing dataset of previously analyzed mortar samples from the Sarno Bath complex. The statistical distribution of the new samples was then compared with the existing data, revealing overlaps within distribution clusters. However, certain samples presented uncertainties that could not be fully addressed by PCA.

To further validate these observations and enhance the attribution of samples to specific construction phases, a discriminant analysis (DA) approach was adopted. This multivariate statistical method, often used for predictive purposes [11], classifies data into categories based on posterior probabilities calculated from observed features. Using a training dataset with pre-classified samples as markers, DA estimates the likelihood that new observations belong to a specific category by considering variable distributions and prior probabilities. In essence, DA compares the position of data points relative to the centers of class distributions while accounting for internal variability, assigning samples to categories that have the highest posterior probability. In archaeology, DA has been applied to relate new observations to classified datasets, particularly in probabilistic determinations comparing archaeological geomaterials (e.g., stone, ceramics) with geologically compatible samples from known sources [12–16]. In this study, DA was innovatively employed as a machine-learning tool for automated phase attribution, based on the compositional compatibility of mortar samples from the 2022 excavation with those previously analyzed and linked to known construction phases. Pre-classified samples from level +0, analyzed in [10], served as training markers. By analyzing the mineralogical affinities of samples, DA assigned the unknown-phase 2022 samples to specific construction phases based on their probabilistic compatibility with known-phase samples, using two confidence intervals.

Both PCA and DA analysis were developed using multivariate statistical tools integrated into Statgraphics Centurion PRO 19 (© 2023 Statgraphics Technologies, Inc., The Plains, VA, USA). The data were ultimately validated through cross-comparison with optical microscopy analysis of samples prepared as 30 μm thin sections. Additionally, SEM-EDS analyses provided insights into the potential sources of the volcanic aggregate fraction further validating the mineralogical differences observed after QPA-XRPD analyses and PCA/DA associations. These findings enhanced the chronological framework of the construction phases and corroborated stratigraphic hypotheses established through field archaeological investigations.

2. The Previous Research

Research on the Sarno Bath complex in Pompeii began in 2016, focusing on a multi-analytical examination of its construction history and state of preservation. This included investigating construction techniques, seismic monitoring, archival data on the building's history, and the materials used, including structural mortars [17].

The Sarno Bath complex, situated in *Regio* VIII, *Insula* 2 near the Forum, represents a significant Roman spa complex constructed on Pompeii's southwestern volcanic plateau [18]. Initially excavated between 1887 and 1890, with additional façade work in the 1950s, the site has a long history of use, dating back to the Archaic and Samnite periods. By the 2nd century BCE, during the Late Republican period, terraced *domus* were built in this area. The baths were realized in a later period and, in their final arrangement, spanned approximately 3700 m^2 across four terraced floors descending the plateau, featuring a façade with an entrance overlooking the Sarno Valley. The ground floor (level +0) comprised three interconnected housing units with independent entrances, today accessible from Vicolo della Regina and Via delle Scuole. Below, level -1 housed cisterns and service rooms were accessed via a vaulted corridor. Level -2 contained three panoramic apartments connected by a decorated cryptoporticus, while level -3 housed the actual baths and seven small rooms of uncertain function along a windowed corridor. Finally, level -4 included five service rooms, above a thick *opus caementicium* foundation.

The multidisciplinary research conducted by the University of Padua provided a deeper understanding of the complex's construction history, enabling the identification of four key phases of its construction and restoration [19].

- Phase I: Isolated structures with unspecified function and chronology identified underneath later buildings of phase II.

- Phase II: Construction of at least two terraced *domus*, including walls and floors and equipped with water cisterns, certainly before the Imperial age.
- Phase III: Reorganization of the *domus* into their current layout likely occurred after the Augustan age (early 1st century CE). This phase involved reframing earlier structures and filling pits that remained incomplete at the time of the 79 CE eruption. A large terraced building was constructed during this phase, housing (or intended to house) the thermal baths on levels -3 and -4, accessible independently via a tunnel leading to Vicolo della Regina.
- Phase IV: 19th- and 20th-century restorations, often poorly documented, altered the original layout and plan of the buildings.

A crucial contribution to the identification of the construction phases came from the analyses of structural mortar samples discussed in [9,10]. Overall, around 80 mortar samples from both dated and undated structures were analyzed using optical microscopy (OM) and X-ray powder diffraction with quantitative phase analysis (QPA-XRPD). Principal component analysis (PCA) of the QPA-XRPD data classified the samples into three main groups, aligned with distinct construction phases and restoration periods. OM validated these groupings by confirming the statistical differentiation of the mineralogical profiles. The groupings attributions are reported hereafter.

➢ Group 1: Modern restoration mortars (phase IV), subdivided into Gr 1a, 1b, and 1c, corresponding to specific restoration interventions. Gr 1c mortars likely date to 19th-century restorations, as supported by thermoluminescence dating of bricks [9]. Gr 1b mortars, characterized by leucite-rich materials, suggest restorations postdating the 1930s.

➢ Group 2: Mortars dated between the end of the Augustan age and 79 CE (phase III), with subgroups Gr 2a and 2b. These likely reflect construction activities before and after the earthquake of 62/63 CE.

➢ Group 3: Mortars attributed to two sub-groups. Subgroup Gr 3a aligns with structures dated to the pre-Imperial age (phase I–II), while for 3b the chronological association is unclear.

PCA analysis identified distinct mineralogical markers differentiating the phases, as reported below. These findings, combined with stratigraphic and archival data, provide a robust chronological framework for understanding the Sarno Baths' architectural evolution before and after the 79 CE eruption of Mt. Vesuvius.

3. The 2022 Excavations at Civ. 21

During the 2022 campaign two excavation trenches were opened at the ground floor (level +0) of the house at civ. 21 of Via delle Scuole. The first one was located in correspondence with rooms G1 and G2 (sector D) and the second one in correspondence with room O (sector E). In both sectors, the excavation activity brought to light structures attributable to the two already known building phases of the complex, namely phase II (pre-Imperial age) and phase III (post-Augustan age). In particular, in sector D it was possible to distinguish some structures and layers certainly belonging to the house currently visible at civ. 21 (phase III) and structures and floors attributable to the earlier building already partially investigated in 2018–2019, called 'Casa A' (phase II). A similar situation was observed in sector E, where both structures related to the *domus* currently visible at civ. 21 (phase III) and evidence referring to a pre-existing *domus* (the so-called 'Casa B', phase) were found.

Regarding chronology, the 2022 excavation finds suggest that the *domus* at civ. 21 was constructed during the 1st century CE and the work was still in progress in 79 CE, while the earlier *domus* (Casa A and Casa B) were likely built in the second half or towards the end of the 2nd century BCE or at the beginning of the 1st century BCE.

4. The 2022 Sampling

In the 2022 research campaign, 15 lime mortar samples were taken and analyzed (Figure 2). They all consisted of bedding mortars of masonry structures (Table 1). Among the wall structures sampled, some were attributed, on a stratigraphic basis, to a specific building phase. This is the case, for example, of walls USM 304 and 302, referable to the so-called 'Casa A' (phase II), and walls USM 81 and 83, clearly part of the more recent *domus* at civ. 21 (phase III). For other structures, the archaeological stratigraphic sequence did not provide clues to identify a possible construction phase. In particular, in some cases, it was unclear whether the entire structure belonged to the same construction phase or if the upper elevation was more recent, built on an older foundation. This is the case, for instance, for walls USM 80, 300, and 301. Given this uncertainty, a sample was taken from both the foundation and the elevation of each structure.

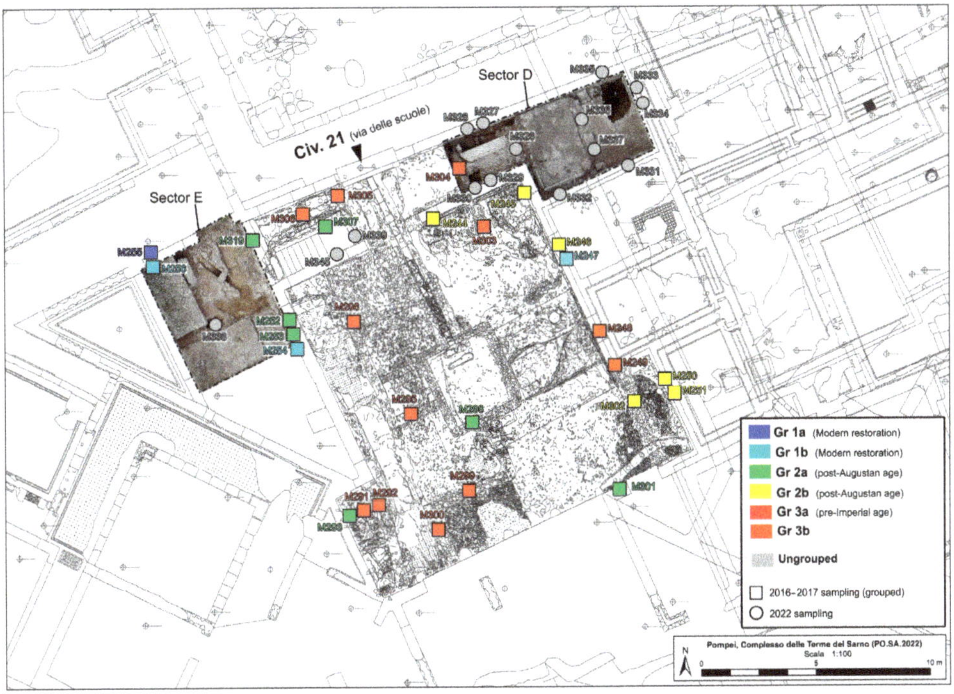

Figure 2. Samples collected from the 2022 excavation (circular labels in grey) together with those already analyzed from civ. 21 and associated to groups identified in [10].

With regard to the construction techniques of the sampled walls, no particular differences were observed. In fact, the foundations all consist of irregular stone elements bonded by lime mortar and arranged in an irregular manner (*opus incertum*). Different lithotypes are often used within the same structure; however, in most cases, compact and vesicular tephrite are most prevalent as masonry elements (Figure 3a). Regarding the elevated portions, two distinct techniques were identified. The first technique involves the use of irregular stone elements of various lithologies (predominantly travertine), arranged in an irregular pattern (*opus incertum*) and bound with lime mortar (Figure 3b). The second technique involves the exclusive use of yellow tuff, shaped into sub-rhomboidal elements arranged in rows, inclined at a 45-degree angle (*opus quasi reticulatum*) (Figure 3c).

Table 1. Samples collected from the 2022 excavation with indication of the structural function, materials, construction technique, and hypothetical building phase (attributed on a stratigraphic basis) of the wall structures sampled.

Sample	Wall Number	Structural Function	Building Materials	Building Technique	Hypothetical Building Phase (Based on Archaeological Data)
M326	USM338	Foundation	Compact and vesicular tephrite	*Opus incertum*	II
M327	USM83	Foundation	Vesicular tephrite, yellow tuff, travertine	*Opus incertum*	III
M328	USM83	Elevation	Yellow tuff	*Opus quasi reticulatum*	III
M329	USM81	Foundation	Compact tephrite	*Opus incertum*	Not determined
M330	USM81	Elevation	Yellow tuff + bricks (in the chains)	*Opus quasi reticulatum*	III
M331	USM301	Foundation	Compact tephrite	*Opus incertum*	Not determined
M332	USM301	Elevation	Yellow tuff	*Opus quasi reticulatum*	III
M333	USM300	Foundation	Vesicular tephrite, white limestone, travertine, bricks	*Opus incertum*	Not determined
M334	USM300	Elevation	Travertine, compact and vesicular tephrite, grey tuff	*Opus incertum*	III?
M335	USM83	Foundation	Vesicular tephrite, yellow tuff, travertine	*Opus incertum*	Not determined
M336	USM304	Foundation	Compact and vesicular tephrite, travertine, bricks	*Opus incertum*	Not determined
M337	USM304	Foundation/ Threshold basis?	Compact and vesicular tephrite, travertine, bricks	*Opus incertum*	Not determined
M338	USM515	Foundation	Travertine	*Opus incertum*	Not determined
M339	USM 173	Elevation	Compact tephrite, reused brick fragments	*Opus incertum*	Not determined
M345	n.d.	Elevation, corner chain	Grey tuff, bricks	*Opus vittatum mixtum*	Not determined

Figure 3. The building techniques used in the wall structures sampled: (**a**) foundation in *opus incertum*; (**b**) elevation in *opus incertum*; (**c**) elevation in *opus quasi reticulatum*; (**d**) elevation in *opus incertum*.

Typically, bricks are used to construct the corner chains in this type of structure. Another documented technique is *opus vittatum mixtum* (Figure 3d), seen in the corner chain of the west wall of the tunnel leading to the lower floor. This technique consists of alternating courses of parallelepiped blocks of grey tuff and bricks.

5. Analytical Workflow and Data Processing

From an analytical perspective, all samples were initially analyzed using QPA-XRPD (see Supplementary Materials S1 for instrumental parameters and equipment details [16,20–22]) to facilitate a mineralogical comparison with the samples reported in [9,10]. Beyond identifying the primary mineral phases in the mortar samples (Figure 4), XRPD analysis also clearly revealed secondary phases associated with alteration processes in some mortars from already exposed structures (raised walls). These included gypsum, halite, and gaylussite, which were linked to the migration and crystallization of soluble salts, as described in [23]. In particular, gypsum was identified as a product of atmospheric weathering on exposed mortars, consistent with previous studies [5,24]. Halite, another common alteration product in ancient mortars, was similarly detected [25]. To ensure consistency in the analysis, the proportions of the remaining primary mineral phases were recalculated to 100% wt (see Supplementary Materials S2).

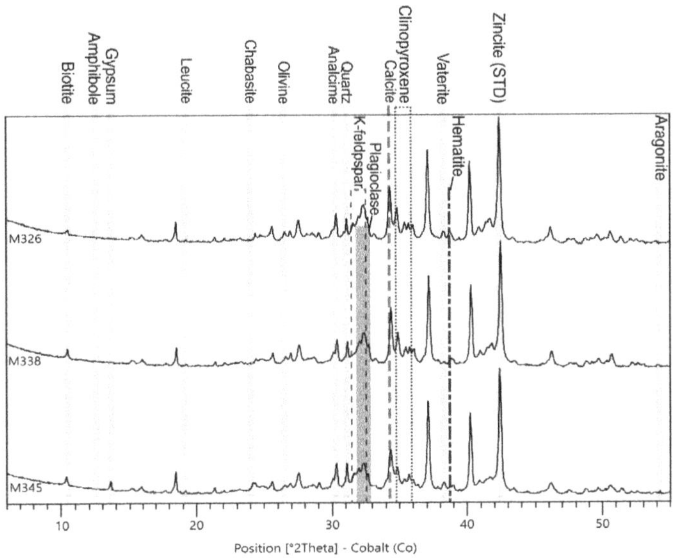

Figure 4. XRPD spectra of representative mortar samples from the 2022 excavation, showing the d-spacing of the principal peak of all mineral phases other than amorphous. For those mineral phases overlapping their main peaks, an indicative secondary peak has been highlighted (i.e., aragonite).

5.1. PCA on Mineralogical Data

The resulting mineralogical data were reprocessed alongside the samples analyzed in [10] from level +0 of the Sarno Baths using PCA. This statistical method extracts a limited number of linear combinations that effectively capture the variability of the original mortar mineralogical profiles. It generates a set of principal components that represent the dataset's variations. This analysis employed the original mineralogical phases of the mortar samples as descriptive variables (recalculated at 100%wt after the removal of the alteration phases). The sample distribution was reported in a scatterplot according to the value of the first two extracted components (PC1, PC2), as the first two components are those describing the main variance of the dataset. The resulting bivariate diagram (Figure 5) highlights strong correlations between the 2022 excavation samples and the previously analyzed samples.

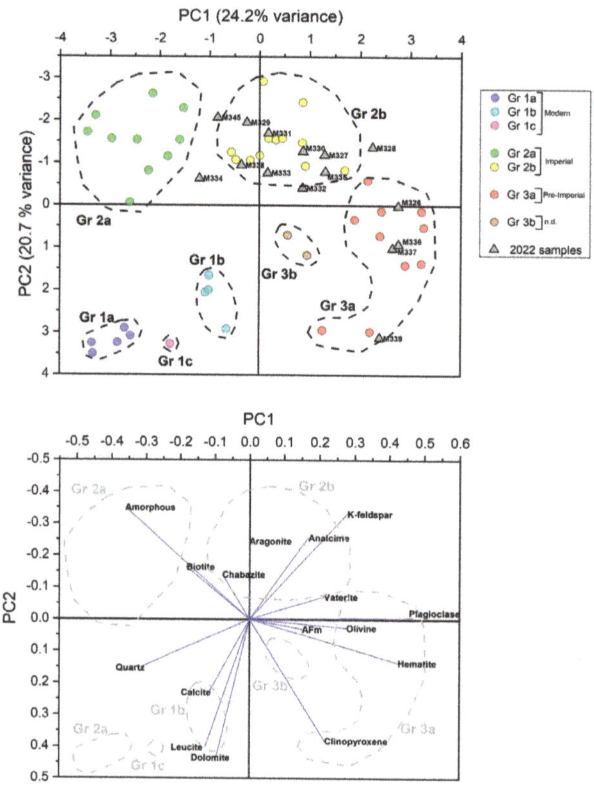

Figure 5. PCA distribution of the 2022 excavation samples according to QPA-XRPD data (triangles), processed together with mortar from level +0, in colored dots according to final groupings identified in [10].

The PCA distribution diagram shows a trend in the PC1 and PC2 axes similar to that observed in [10], with vectors indicating mineralogical phases that are interrelated and characteristic of the identified sample groups:

- Group 3a (PC1 and PC2 values between 0 and +4). This group comprises samples attributed to the pre-Imperial construction period. The connoting mineral phases, positively correlated with group distribution, include hematite, plagioclase, olivine, and, subordinately, clinopyroxene (Cpx). Some 2022 samples, such as M339, M336, M337, and M326, fall within this range.
- Group 3b (PC1 and PC2 ~+1). No 2022 excavation samples clearly align with this subgroup.
- Group 2a (PC1: −1 to −4, PC2: 0 to −3). These samples, dating to the pre-79 CE Imperial phase, are characterized by high concentrations of amorphous phases and biotite. None of the 2022 samples fall directly within this interval. However, samples M334 and M345 lie in intermediate positions between groups 2a and 2b.
- Group 2b (PC1: −1 to +2, PC2: 0 to −3): This group also collects samples from the pre-79 CE Imperial phase. This group is defined by positively correlated phases such as K-feldspar, analcime, and, to a lesser extent, aragonite. Most 2022 samples, including M327, M329, M330, M331, M333, M335, and M338, and, align with this group. Petrographic analysis reveals a higher concentration of pyroxenes (augite), olivine, and volcanic clasts with K-feldspar microlites (sanidine) compared to group 2a mortars.

- Modern restoration mortars (groups 1a, 1b, and 1c, PC1: 0 to −4, PC2: 0 to +4): These groups are defined by positive correlations with calcite, leucite, dolomite, and quartz. None of the 2022 samples fall within or near this quadrant, suggesting no modern restoration mortars were sampled.

From the PCA results, even if it is clear that no 2022 samples correspond to modern restoration mortars, certain correlations between some samples and specific groups remain ambiguous. In fact, several samples occupy intermediate positions, particularly between groups 3a and 2a/2b. This suggests that, while the 2022 samples align with established construction phases, mineralogical fingerprints alone may not definitively associate samples with specific groups. This is partly due to the inherent limitations of PCA analysis, as the first two PCs together account for only about 45% of the total dataset's variability. Secondary correlations can be explained by analyzing the remaining PCs with eigenvalues greater than or equal to 1.0 (PC3 to 6), which, however, account for minor fractions of the dataset's variability (each contributing less than 10%).

5.2. Discriminant Analysis of Mineralogical Data

Discriminant analysis was employed as a machine-learning predictive method to automatically attribute samples of uncertain phase to a specific construction phase. The training dataset comprised the same samples used in the PCA analysis, representing structures already assigned to construction phases of level +0 of the Sarno Bath complex. Mineralogical phases descriptive of the dataset served as independent discriminant variables, while the 2022 excavation samples were treated as uncertain data to be probabilistically associated with a construction phase based on their compatibility with the centroids of the reference sample groups. A total of 45 cases were used to develop a model to discriminate among the six levels of attribution. Using a stepwise selection algorithm, it was determined that six variables (amorphous, analcime, hematite, olivine, leucite, dolomite) were significant predictors of group. The four discriminating functions with p-values less than 0.05 are statistically significant at the 95.0% confidence level. The data processing enabled the assignment of most samples to a specific group (i.e., construction phase) with high confidence (>99% for 1st probability), as reported in Table 2.

Table 2. Results of the discriminant analysis (DA, stepwise method—forward selection) on the samples from the 2022 excavation attributed to groups at two probabilistic intervals.

Sample	Assigned Group (1st)	Prob. (%)	Assigned Group (2nd)	Prob. (%)	Based on PCA
M326	3a	100	*null*	*null*	3a
M336	3a	100	*null*	*null*	3a
M337	3a	100	*null*	*null*	3a
M339	3a	100	*null*	*null*	3a
M327	2b	100	*null*	*null*	2b
M328	2b	100	*null*	*null*	2b ?
M329	2b	99.87	2a	0.13	2b
M330	2b	99.03	3b	0.97	2b
M331	2b	99.98	2a	0.02	2b
M333	2b	99.98	3b	0.02	2b
M334	2b	99.41	2a	0.59	2a or 2b
M335	2b	100	*null*	*null*	2b
M338	2b	99.06	2a	0.94	2b
M345	2a	88.37	2b	11.63	2a or 2b
M332	3b	51.85	2b	48.15	2b ?

- Group 3a: Samples M326, M336, M337, and M339 were confidently assigned to this group, consistent with the PCA results.
- Group 2b: Samples M327, M328, M329, M330, M331, M333, M334, M335, and M338 were reliably attributed to this group.
- Group 2a: Sample M345 was assigned to group 2a with good probability (88.37% 1st probability); however, a potential association with group 2b cannot be excluded (11.63% 2nd probability), highlighting the difficulty in definitively attributing this sample to one of these groups.
- Sample M332: This sample's attribution remains uncertain. DA indicated nearly equal probabilities of belonging to group 3b or group 2b. This uncertainty is possibly due to the limited representativeness of the reference dataset, which lacks the larger, categorized markers (e.g., phased mortars) typically used in DA. Expanding the dataset with more reference markers could significantly reduce this uncertainty. Based on PCA observations of the first two principal components, this sample seems more compatible with group 2b.

5.3. OM and SEM-EDS Analyses (Data Verification)

The results of the statistical processing of the mineralogical descriptive data were subsequently validated through petrographic and chemical characterization of the compounds, using transmitted light optical microscopy and SEM-EDS investigations of representative samples prepared in 30 μm thin sections (see Supplementary Materials S1 for details on instrumental parameters and equipment [16,20–22]) This analysis further confirmed the attribution of the samples based on the distinctive petro-mineralogical features that characterize the mortars sampled in the 2022 archaeological campaign. The samples attributed to group 3a (M326, M336, M337, and M339) are primarily composed of highly vesicular tephrites exhibiting a hypocrystalline hyalopilitic structure. These aggregates contain scattered phenocrysts, typically constituted by clinopyroxenes (augite–diopside) and feldspars, predominantly of the plagioclase variety (labradorite/bytownite), with subordinate K-feldspars (sanidine), often arranged as glomerophyres (Figure 6a). Moreover, single-grained leucite phenocrysts are usually almost completely transformed into analcime due to analcimization processes [26] (Figure 7(a–b2)). Accessory minerals include olivine, biotite, hematite, and opaque minerals (Figure 8(a1–a3)). The vesicular tephrite component ranges from multi-millimeter-sized clasts to powdered fractions, forming cineritic shards tightly blended with the lime-based binder. The binder is carbonate-based, homogeneous, micritic, and thoroughly integrated with such scoriaceous aggregate, typically in binder-to-aggregate (B:A) ratios of approximately 1:2 or 1:3. Unmixed lime lumps are minimal. The mortar porosity is low, with rounded vesicle-type pores predominating. A completely subordinate fraction of the aggregate is constituted by extremely scattered grains of pyroclastic rocks (pumices).

The samples confidently attributed to group 2b (M327, M328, M329, M330, M331, M333, M334, M335, and M338) include the same highly vesicular tephrites found in group 3a (Figure 6b). However, in the mortars of group 2b, these tephrites are mixed in comparable proportions with pumice clasts exhibiting reacted borders. They can be classified into two main types: one is highly vesicular and aphyric/vitrophyric, with scattered biotite phenocrysts (Figure 8(b1–b3)), while the other is feebly porphyritic/hyalopilitic with reduced internal vascularity (Figure 8(c1–c3)). Rarely, scattered tuffs with cineritic matrices were identified. They embed pumices, volcanic shards, and individual mineral components such as biotite, K-feldspar, occasional clinopyroxenes, and opaque minerals. A subordinate component of the aggregate includes grains of effusive rocks with porphyritic/trachytic textures, characterized by feldspars with sanidine > plagioclase > anorthoclase, and opaque minerals, along with accessory clinopyroxenes and biotite. The carbonate binder in these samples is micritic to microsparitic, slightly heterogeneous, with a greater occurrence of poorly mixed lime lumps. The binder-to-aggregate ratio (L:A) tends towards 1:2. The open porosity is scarce and primarily constituted by vugh/vesicle type voids.

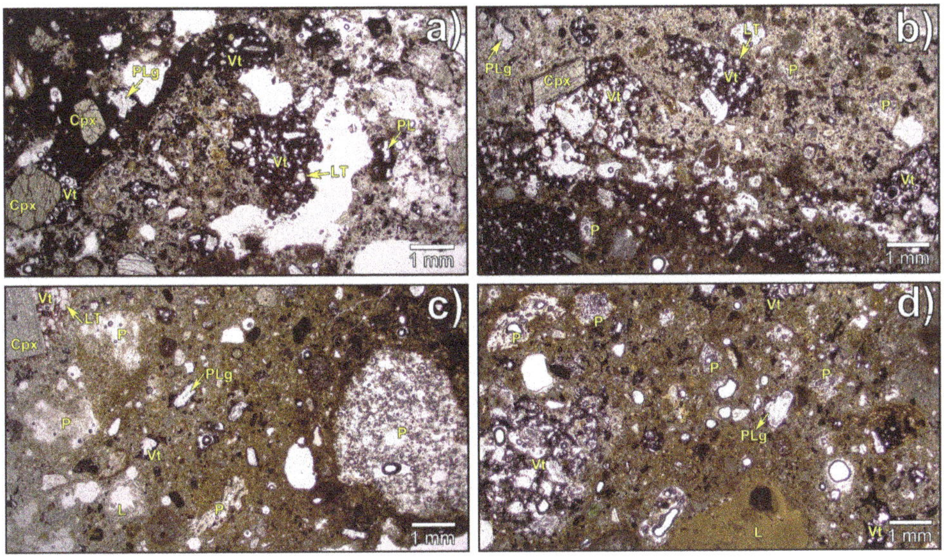

Figure 6. Micrographs in optical microscopy with transmitted polarized light (parallel nicol) of representative samples from the groups previously defined. (**a**) M326 (group 3a); (**b**) M338 (group 2b); (**c**) M345 (group 2a); (**d**) M332 (compatible with group 2b). Legend: Vt = vesicular tephrite; *p* = pumice; PL = plagioclase; PLg = plagioclase (glomerophyre); L = lump; LT = leucite (usually analcimized); Cpx = clinopyroxene.

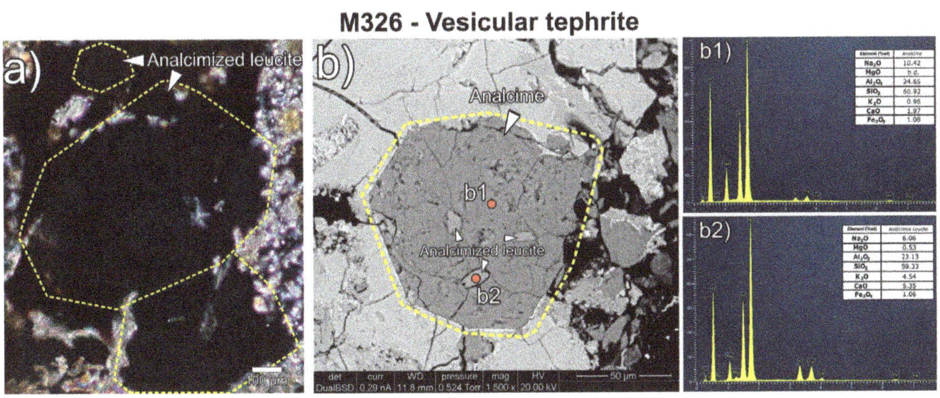

Figure 7. Analcimized leucite mineralization in a vesicular tephrite clast included in sample M326. (**a**) TL-OM cross-polar high-magnification micrograph of analcimized leucite. (**b**) SEM image (backscattered electron) of the analcimized leucite; (**b1**,**b2**) EDS microanalyses in two points of the analcimized leucite.

Sample M345 differs from the group 2b samples as its aggregate fraction is primarily composed of aphyric juveniles with scattered biotite phenocrysts, and slightly porphyritic/hyalopilitic pumices. The presence of grains and shards of vesicular tephrite

(Figure 6c) is significantly lower. Based on these characteristics, it may be classified with the group 2a mortars, which are connoted by a high concentration of amorphous material and biotite, quantified also after QPA-XRPD analyses, indicative of the elevated pumice fraction included in the sample constituting this group. A totally subordinate component of the aggregates consists of more compact tephrite grains. These grains often contain phenocrysts of biotite, clinopyroxene, plagioclase, and minerals of leucite, with a lower degree of analcimization (Figure 9(a–a2)).

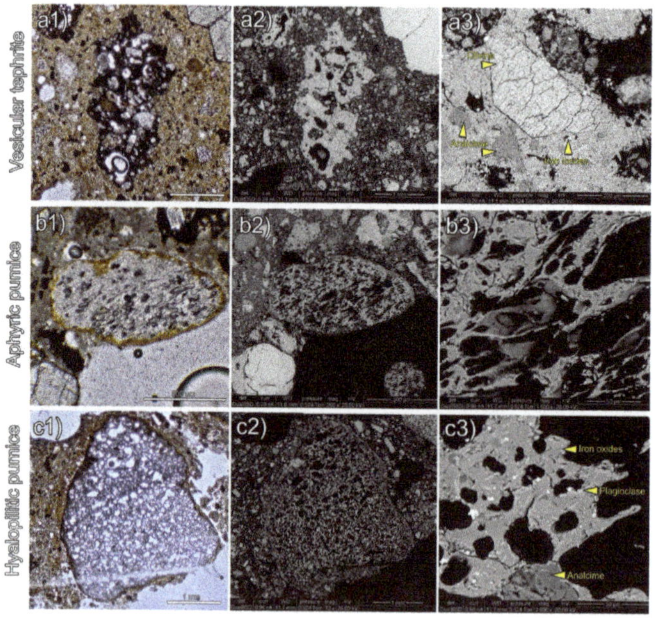

Figure 8. OM-TL (parallel nicols) and SEM (backscattered electrons) acquisition of representative volcanic clasts included in mortars. (**a1–a3**) vesicular tephrite; (**b1–b3**) aphyric pumice; (**c1–c3**) hyalopilitic pumice.

Finally, although sample M332 is probabilistically associated by DA to group 2b or group 3b (with nearly equal likelihood), petrographic analysis suggests it does not exhibit significant differences from the group 2b samples (Figure 6d). Thus, it is more plausibly associated with group 2b, aligning with the observations from the PCA distributions. The mortar features observed via OM coupled with SEM-EDS support the PCA distribution of samples according to their QPA-XRPD profiles.

In conclusion, samples from group 3a are positively correlated with trend vectors for plagioclase, hematite, clinopyroxene, and olivine, indicative of the common mineralizations detected in the high tephritic/cineritic fraction present in these composites. In contrast, group 2a samples exhibit an inverse correlation in respect to the mineralogical phases that characterize group 3a mortars. In fact, group 2a mortars feature positive correlations of amorphous fraction and biotite content, which reflects the elevated presence of aphyric pumice in the mortars of these composites. Intermediate correlations are observed in group 2b samples, which show a high concentration of analcime, attributable to the analcimized leucites found in both vesicular tephrites and certain pumices. These mineral phase vectors, in turn, are entirely counter-correlated with samples from modern restorations, distinguished by positive correlations with calcite content (indicative of lime-rich mortars), quartz (different source of certain aggregates), and leucite [10]. In particular, the high leucite content in the group 1 mortars is likely associated with the presence of pumice fragments enriched with leucite, characterized by abundant mineralizations within the volcanic glass unaffected by analcimization, as detected in modern restoration sample M259 analyzed in [10] (Figure 10).

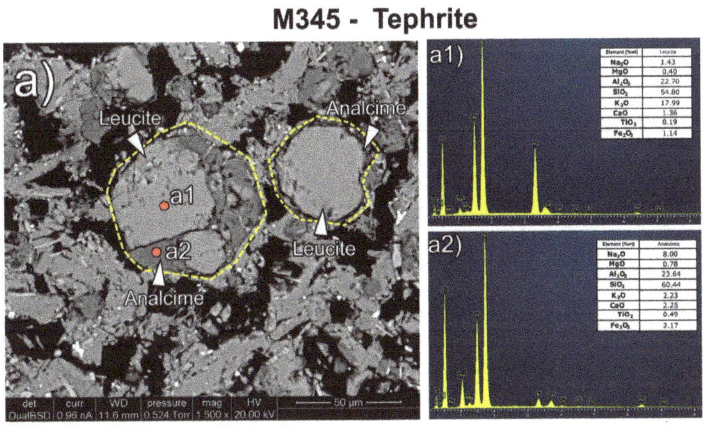

Figure 9. Leucite mineralization in a tephrite included in sample M345. (**a**) SEM image (backscattered electron) of the leucite mineral with clear analcimization rims; (**a1**,**a2**) EDS microanalyses in two points of the analcimized leucite.

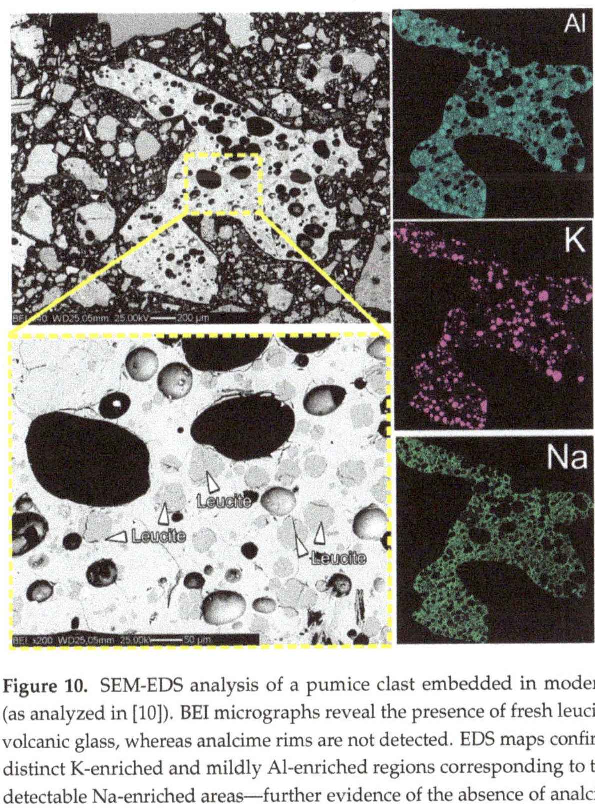

Figure 10. SEM-EDS analysis of a pumice clast embedded in modern restoration mortar M259 (as analyzed in [10]). BEI micrographs reveal the presence of fresh leucite mineralization within the volcanic glass, whereas analcime rims are not detected. EDS maps confirm this observation, showing distinct K-enriched and mildly Al-enriched regions corresponding to the leucite minerals, with no detectable Na-enriched areas—further evidence of the absence of analcimization processes.

6. Discussion

6.1. Sequencing Building Interventions

The analyses of the 2022 mortars allowed the sampled structures to be attributed with certainty to a specific building phase based on the characteristics of the mortar employed (Figure 11). In some cases, the analytical results confirmed the hypotheses already proposed on an archaeological–stratigraphic basis. In other instances, the analysis resolved uncertainties regarding the attribution of certain structures to specific building phases. Notably, the absence of differences between the mortars used in the foundations and those in the elevations clarified that certain structures (USM 80, 300, and 301) were not constructed atop older walls but were entirely built within the same phase (phase III) from the foundation upward. This insight enabled a more accurate interpretation of the building's architectural evolution and construction processes.

The analysis also provided critical insights for refining the sequence of construction activities within the complex. Notably, the classification of sample M338 within group 2b revealed an important revision of prior assumptions [10]. In fact, it demonstrated that mortars from group 2b actually predate those of group 2a. This conclusion is supported by the fact that the sample originates from a levelled wall structure unequivocally associated with an earlier construction phase (USM 515), predating the house currently visible at civ. n. 21. Indeed, this structure is overlaid by the kitchen counter in room O, likely realized in the last phase of restoration before the 79 CE eruption (Figure 12).

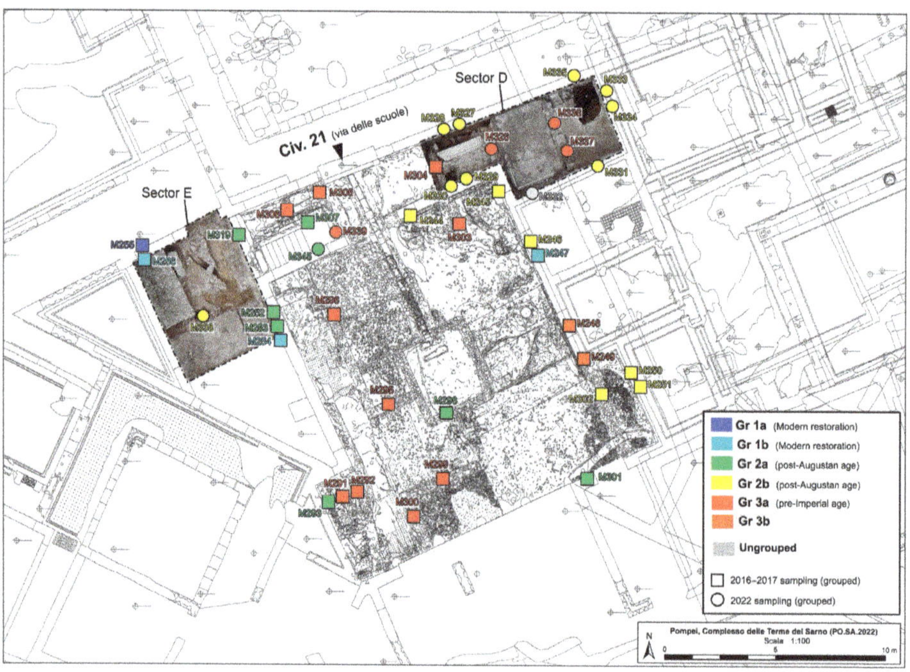

Figure 11. Samples collected from the 2022 excavation in circular labels colored according to attributed groups after the processing of analytical data.

As for the building techniques, the mortar analysis confirmed what had already been hypothesized on the basis of archaeological data, namely that in the complex there is no close relationship between construction techniques and building phases, in particular in the more recent one. In fact, even if *opus incertum* seems to predominate in phase II structures (characterized by group 3a mortars), at least at foundation level, the same technique continued to be used in phase III structures, in which other techniques are also recognizable (*opus quasi reticulatum*, *opus vittatum mixtum*). A similar trend can also be observed with regard to the materials used in the wall structures. In phase II walls, compact and vesicular tephrite predominate (though not exclusively), while in phase III structures, a wider variety of stone materials is found (compact and vesicular tephrite, but also travertine, yellow tuff, and grey tuff), sometimes used in combination with brick.

Figure 12. The wall USM 515 overlaid by the kitchen counter in room O (sector E).

6.2. Chronology and Raw Materials

The aspects outlined above allow for an analysis of the formulation of mortars used in the building, identifying a clear evolutionary trend in the combinations of raw materials, particularly concerning the aggregates. Petrographic investigations clearly reveal a significant evolution in mortar production, especially in relation to the aggregate fraction. In the samples of group Gr 3a, attributed to phase II and associated with the earliest construction periods of the *domus* in the Sarno Bath complex, there is a predominant presence of ground vesicular tephrite, milled down into fragments or powder. These soft stones, also known as *Lava Foam* or *Lava-Schiuma* [27–29], were extensively quarried and employed in construction in Pompeii [29–32]. In the Sarno Baths, they constitute a recurrent material adopted for masonry structures [33]; in fact, this stone type was employed in the construction of phase II walls investigated along with the 2022 excavations in the civ. 21 (see Table 1). Discarded chips during the in situ cutting of these soft stones were subsequently repurposed as aggregate in the mortar production for phases I and II. According to the TAS diagram (Figure 13a), these clasts exhibit a chemical composition ranging from basaltic trachyandesite (approaching shoshonite) to tephriphonolite, aligning perfectly with the chemical range of Somma–Vesuvius lavas [34,35] (for the results of the SEM-EDS analyses on volcanic aggregates, see Supplementary Materials S3).

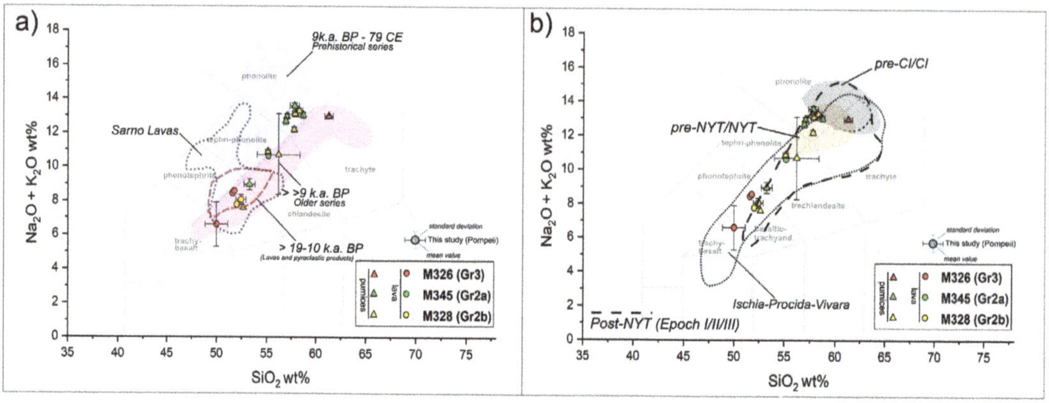

Figure 13. TAS scatterplots of pumices clasts and lavas (tephrites) included in samples M326 (Gr 3a), M345 (Gr 2a), and M328 (Gr 2b), respectively, overlapped to the geochemical intervals of compatible volcanic products of the Neapolitan district: (**a**) three primary eruptive facies of Somma–Vesuvius (reference from [22] and related sources; lava geochemical intervals are modified from [30]); (**b**) main eruptive events of the Phlegraean Fields, comprising pre-Campanian and Campanian Ignimbrite (pre-CI/CI), pre-Neapolitan and Neapolitan Yellow Tuff (pre-NYT/NYT), post-NYT, and Phlegraean-correlated volcanoes of Ischia and Procida–Vivara (reference from [22] and related sources).

In the mortars of group Gr 2b, we detected a significant increase in pumice grains, which become predominant in group Gr 2a, although vesicular tephrite lavas remain present to some extent. Moreover, silica-undersaturated pumice clasts, which range in composition from tephriphonolite to basaltic trachyandesite, are less common. In contrast, most of the juveniles, and in particular the aphyric clasts, primarily exhibit a phonolitic chemical composition. This geochemical composition overlaps with the main Phlegraean volcanic products [36,37] (Figure 13b) as well as the Somma–Vesuvius older (>9 k.a. BP) and prehistorical (9 k.a. BP—79 CE) Plinian and sub-Plinian explosive series [38,39] (Figure 13a). Consistent with the findings of [8] regarding the clasts present in Pompeian mortars from the Samnite period, our analysis of the Gr 3a mortars, associated with pre-Imperial construction phases, similarly outlines the absence of pumices with phonolitic compositions. Moreover, it is noteworthy that such chemical compositions are often identified in mortars through the analysis of pumices serving as pozzolanic aggregates, where these clasts are associated with Phlegraean eruptions [16,22,40], particularly those occurring after the Neapolitan Yellow Tuff event (post-NYT eruptions). This aspect aligns well with the increasing use of yellow tuff in masonry, likely linked to Phlegraean volcanic activity [36,37], particularly during the pre- and post-earthquake renovations of 62/63 CE. The first evidence of its use extends back to the late Republican period and continues to expand into later decades [32,41,42]. The increasing use of yellow tuff as masonry material in the construction and reconstruction activities of the Imperial period in Pompeii could therefore be associated with the growing presence of non-lithified pyroclastic debris, such as pumice and aphyric shards, which were used as pozzolanic aggregates in mortars and likely sourced predominantly from Phlegraean-related deposits.

7. Conclusive Remarks

The multi-analytical protocol we adopted, primarily based on the processing of mineralogical QPA-XRPD data using multivariate statistical analysis techniques, proved highly effective in distinguishing construction phases and identifying raw materials used in the Sarno Bath complex. This effectiveness is contingent on the availability of a comprehensive reference dataset of samples associated with well-defined phases. The dataset must encompass all potential phases and subphases to ensure reliable classification of unknown samples. The method we adopted acts like a machine-learning tool, efficiently assigning samples to groups based on a well-classified training dataset. This semi-automated protocol is both cost effective and time efficient, addressing group classification challenges with high confidence by maximizing the use of the mineralogical characteristics of the samples. Ambiguous cases, identified by low probabilistic attribution through discriminant analysis, are flagged for further verification using complementary techniques such as optical microscopy and SEM-EDS. Cross-validation of data and comparison with archaeological results has demonstrated the method's effectiveness. Following this data processing, the samples still remaining uncertain are limited to a few instances (e.g., M332). In such cases, it may be necessary to employ other advanced techniques (e.g., XRF on bulk mortars, or FT-IR) to more confidently define their group affiliations or confirm their status as outliers.

In conclusion, this method can streamline workflows and reduce subjective and speculative interpretations and assumption. Therefore, it could be effectively applied to the study of other ancient buildings with long construction histories, not limited to Vesuvian sites. However, it is crucial to first establish a comprehensive reference dataset that captures the variability of ancient and modern mortars with a well-defined chronology. Such a dataset can then be used to attribute other structures of uncertain chronology to specific construction or restoration episodes.

Finally, this analytical protocol can significantly aid structural investigations for diagnostic purposes and maintenance or restoration of buildings by distinguishing sectors constructed at different times, which therefore may present diverse conservation risks. From an archaeological perspective, it provides a cost-effective alternative to the extensive and time-consuming stratigraphic excavation traditionally needed to establish the chronology of structures based on associated artifact assemblages and absolute dating methods (i.e., ^{14}C), or alternatively serves as an efficient pre-screening tool for guiding targeted and spatially limited stratigraphic investigations.

Supplementary Materials: The following supporting information can be downloaded at: https://www.mdpi.com/article/10.3390/heritage8020051/s1, S1: Instrumental equipment and standards; S2: Results of the QPA-XRPD analyses; S3: Results of the SEM-EDS analyses on volcanic aggregates.

Author Contributions: Conceptualization S.D.; methodology, S.D., M.S.; validation, S.D., C.P.; formal analysis, S.D.; investigation, S.D., C.P.; resources, M.S.B.; data curation, S.D., C.P.; writing—original draft preparation, S.D. (Sections 1, 2, 4, 5, 6.2, 6.3, 7), C.P. (4, 6.1), M.S.B. (3); writing—review and editing, S.D., C.P., M.S., M.S.B.; visualization, S.D., C.P.; funding acquisition, S.D., M.S.B. All authors have read and agreed to the published version of the manuscript.

Funding: This research was funded by the project "Trade and use of volcanic pozzolans in the Roman world. A natural material for the production of eco-sustainable concrete of antiquity" (Principal investigator: Simone Dilaria, BIRD 2023 of the Department of Cultural Heritage of the University of Padua, project code BIRD230232/23). The research infrastructures were implemented and funded within the scopes of the University of Padua under the World Class Research Infrastructures (WCRI) program—SYCURI (Synergic Strategies for Culture Heritage at Risk).

Data Availability Statement: The data presented in this study are available on request from the corresponding authors.

Acknowledgments: The complex has been the subject of study by the University of Padua since 2016 by virtue of an initial agreement signed initially with the then Special Superintendence of Pompeii (prot. STPD11B3LB), then with the Archaeological Park of Pompeii (Rep. no. 71/2020, prot. no. 1310 of 4 August 2020). We would like to thank M. Osanna, G. Zuchtriegel, and the successive officials of *Regio* VIII (S. Masseroli, L. D'Esposito, F. Muscolino and the current G. Scarpati) for their helpfulness and fruitful cooperation.

Conflicts of Interest: The authors declare no conflicts of interest.

References

1. Piovesan, R.; Curti, E.; Grifa, C.; Maritan, L.; Mazzoli, C. Ancient plaster technology: Petrographic and microstratigraphic analysis of plaster-based building materials from the Temple of Venus, Pompeii. In *Interpreting Silent Artefacts: Petrographic Approaches to Archaeological Ceramics*; Quinn, P.S., Ed.; Archaeopress: Oxford, UK, 2009; pp. 65–79.
2. Freccero, A. *Pompeian Plasters, Insula I 9 and Forum*; The Swedish Institute in Rome. Projects and Seminars 5:1; Swedish Institute in Rome Publication: Rome, Italy, 2005.
3. Bonazzi, A.; Santoro, S.; Mastrobattista, E. Caratterizzazione archeometrica delle malte e degli intonaci dell'Insula del Centenario. In *Pompei. Insula del Centenario (IX, 8) I. Indagini Diagnostiche Geofisiche e Analisi Archeometriche*; Santoro, S., Ed.; Studi e Scavi. Nuova Serie 16; Ante Quem: Bologna, Italy, 2007; pp. 93–128.
4. Demauro, T. Lo studio delle malte. Strumento di verifica. In *Restauri a Pompei (1748–1860)*; Demauro, T., Ed.; Studi e Ricerche del Parco Archeologico di Pompei 44; «L'Erma» di Bretschneider: Roma, Italy, 2020; pp. 143–161.
5. Miriello, D.; Barca, D.; Bloise, A.; Ciarallo, A.; Crisci, G.M.; De Rose, T.; Gattuso, C.; Gazineo, F.; La Russa, M.F. Characterisation of archaeological mortars from Pompeii (Campania, Italy) and identification of construction phases by compositional data analysis. *J. Archaeol. Sci.* **2010**, *37*, 2207–2223. [CrossRef]
6. Miriello, D.; Bloise, A.; Crisci, G.M.; De Luca, R.; De Nigris, B.; Martellone, A.; Osanna, M.; Pace, R.; Pecci, A.; Ruggieri, N. New compositional data on ancient mortars and plasters from Pompeii (Campania—Southern Italy): Archaeometric results and considerations about their time evolution. *Mater. Charact.* **2018**, *146*, 189–203. [CrossRef]
7. De Luca, R.; Miriello, D.; Pecci, A.; Domínguez-Bella, S.; Bernal-Casasola, D.; Cottica, D.; Bloise, A.; Crisci, G.M. Archaeometric Study of Mortars from the Garum Shop at Pompeii, Campania, Italy. *Geoarchaeology* **2015**, *30*, 330–351. [CrossRef]
8. Franceschini, M.M.N.; Casa, G.; Calandra, S.; Ismaelli, T.; Grifa, C.; Mercurio, M.; Amoretti, V.; Zuchtriegel, G.; Cantisani, E. Raw materials and building technologies in the public buildings of Pompeii after the earthquake of 62/63 CE: A diachronic analysis of mortars. *Case Stud. Constr. Mater.* **2024**, *21*, e03943. [CrossRef]
9. Secco, M.; Previato, C.; Addis, A.; Zago, G.; Kamsteeg, A.; Dilaria, S.; Canovaro, C.; Artioli, G.; Bonetto, J. Mineralogical clustering of the structural mortars from the Sarno Baths, Pompeii: A tool to interpret construction techniques and relative chronologies. *J. Cult. Herit.* **2019**, *40*, 265–273. [CrossRef]
10. Dilaria, S.; Previato, C.; Secco, M.; Busana, M.S.; Bonetto, J.; Cappellato, J.; Ricci, G.; Artioli, G.; Tan, P. Phasing the history of ancient buildings through PCA on mortars' mineralogical profiles: The example of the Sarno Baths (Pompeii). *Archaeometry* **2022**, *64*, 866–882. [CrossRef]
11. Boedeker, P.; Kearns, N.T. Linear Discriminant Analysis for Prediction of Group Membership: A User-Friendly Primer. *Adv. Methods Pract. Psychol. Sci.* **2019**, *2*, 250–263. [CrossRef]
12. Montana, G.; Cau Ontiveros, M.Á.; Polito, A.M.; Azzaro, E. Characterisation of clayey raw materials for ceramic manufacture in ancient Sicily. *Appl. Clay Sci.* **2011**, *53*, 476–488. [CrossRef]
13. Schürch, B.; Wettengl, S.; Fröhle, S.; Conard, N.; Schmidt, P. The origin of chert in the Aurignacian of Vogelherd Cave investigated by infrared spectroscopy. *PLoS ONE* **2022**, *17*, e0272988. [CrossRef]
14. Prochaska, W. The use of geochemical methods to pinpoint the origin of ancient white marbles. *Mineral. Petrol.* **2023**, *117*, 401–409. [CrossRef]
15. Dilaria, S.; Bonetto, J.; Germinario, L.; Previato, C.; Girotto, C.; Mazzoli, C. The stone artifacts of the National Archaeological Museum of Adria (Rovigo, Italy): A noteworthy example of heterogeneity. *Archaeol. Anthr. Sci.* **2024**, *16*, 14. [CrossRef]
16. Dilaria, S.; Ricci, G.; Secco, M.; Beltrame, C.; Costa, E.; Giovanardi, T.; Bonetto, J.; Artioli, G. Vitruvian binders in Venice: First evidence of Phlegraean pozzolans in an underwater Roman construction in the Venice Lagoon. *PLoS ONE* **2024**, *19*, e0313917. [CrossRef] [PubMed]
17. Artioli, G.; Ghedini, E.F.; Modena, C.; Bonetto, J.; Busana, M.S. Foreword: The MACH Project and the case study of the Sarno Baths in Pompeii. *J. Cult. Herit.* **2019**, *40*, 228. [CrossRef]
18. Bernardi, L.; Busana, M.S. The Sarno Baths in Pompeii: Context and state of the art. *J. Cult. Herit.* **2019**, *40*, 231–239. [CrossRef]
19. Busana, M.S.; Andreatta, C. Abitare a Pompei 'Vista Mare': Il Caso del Complesso delle Terme del Sarno (VIII 2, 17–23). *Atlante Temat. Topogr. Antica 36* **2019**, *in press*.
20. Rietveld, H.M. Line Profiles of Neutron Powder-diffraction Peaks for Structure Refinement. *Acta Crystallogr.* **1967**, *22*, 151–152. [CrossRef]
21. Dollase, W. Correction of Intensities for Preferred Orientation in Powder Diffractometry: Application of the March Model. *J. Appl. Crystallogr.* **1986**, *19*, 267–272. [CrossRef]
22. Dilaria, S.; Ghiotto, A.R.; Secco, M.; Furlan, G.; Giovanardi, T.; Zorzi, F.; Bonetto, J. Early exploitation of Neapolitan pozzolan (pulvis puteolana) in the Roman theatre of Aquileia, Northern Italy. *Sci. Rep.* **2023**, *13*, 4110. [CrossRef]
23. Melander, J.M.; Lauersdorf, L.R. *Masonry: Design and Construction, Problems and Repair*; ASTM: Philadelphia, PA, USA, 1993.
24. Leone, G.; De Vita, A.; Magnani, A.; Rossi, C. Characterization of archaeological mortars from Herculaneum. *Thermochim. Acta* **2016**, *624*, 86–94. [CrossRef]

25. Fahmy, A.; Molina-Piernas, E.; Martínez-López, J.; Domínguez-Bella, S. Salt weathering impact on Nero/Ramses II Temple at El-Ashmonein archaeological site (Hermopolis Magna), Egypt. *Herit. Sci.* **2022**, *10*, 125. [CrossRef]
26. Giampaolo, C.; Godano, R.F.; Di Sabatino, B.; Barrese, E. The alteration of leucite-bearing rocks: A possible mechanism. *Eur. J. Miner.* **1997**, *9*, 1293–1310. [CrossRef]
27. Cinque, A.; Irollo, G. Il "vulcano di Pompei": Nuovi dati geomorfologici e stratigrafici. *Quat. Ital. J. Quat. Sci.* **2004**, *17*, 101–116.
28. Di Girolamo, P. Un esempio di lava schiuma (foam lava) in Campania (Lava schiuma di Pompei Scavi). *Rend. Accad. Sci. Fis. Mat. (Sez. Soc. R. Napoli) Napoli* **1968**, *4*, 4–12.
29. Frizot, M. Caracterization des materiaux de construction de Pompei. In *Dégradation et Restauration de l'architecture Pompéienne*; Adam, J.-P., Frizot, M., Eds.; Éditions du Centre National de la Recherche Scientifique: Paris, France, 1983; pp. 32–41.
30. Kastenmeier, P.; Di Maio, G.; Balassone, G.; Boni, M.; Joachimski, M.; Mondillo, N. The source of stone building materials from the Pompeii archaeological area and its surroundings. *Period. Mineral.* **2010**, *79*, 39–58. [CrossRef]
31. Dessales, H. Les savoir-faire des maçons romains, entre connaissance technique et disponibilité des matériaux. Le cas pompéien. In *Les Savoirs Professionnels des Gens de Métier. Études Sur le Monde du Travail Dans les Sociétés Urbaines de l'empire Romain*; Monteix, N., Tran, N., Eds.; Centre Jean Bérard: Napoli, Italy, 2011; pp. 41–63.
32. Covolan, M. *Tufo Gallo: Cantieri di Costruzione ed Economia del Tufo a Pompei*; Costruire nel Mondo Antico 6; Quasar: Roma, Italy, 2023.
33. Piovesan, R.; Maritan, L.; Meneghin, G.; Previato, C.; Baklouti, S.; Sassi, R.; Mazzoli, C. Stones of the façade of the Sarno Baths, Pompeii: A mindful construction choice. *J. Cult. Herit.* **2019**, *40*, 255–264. [CrossRef]
34. Santacroce, R.; Cioni, R.; Marianelli, P.; Sbrana, A. Understanding Vesuvius and preparing for its next eruption. In *Cultural Responses to the Volcanic Landscape: The Mediterranean and Beyond*; Balmuth, M.S., Chester, D.K., Johnston, P.A., Eds.; Colloquia and Conference Papers 8; Archaeological Institute of America: Boston, MA, USA, 2007; pp. 27–55.
35. Guarino, V.; Solone, R.; Casalini, M.; Franciosi, L.; Dallai, L.; Morra, V.; Conticelli, S.; Melluso, L. The geochemistry of leucite-bearing lavas from early stages of the Somma-Vesuvius volcanic complex: Feeder systems and mantle enrichment processes in the Neapolitan district of the Roman Magmatic Province. *Geochemistry* **2024**, *84*, 126076. [CrossRef]
36. Peccerillo, A. Campania volcanoes: Petrology, geochemistry, and geodynamic significance. In *Vesuvius, Campi Flegrei, and Campanian Volcanism*; De Vivo, B., Belkin, H.E., Rolandi, G., Eds.; Elsevier: Amsterdam, The Netherlands, 2020; pp. 79–120. [CrossRef]
37. Morra, V.; Calcaterra, D.; Cappelletti, P.; Colella, A.; Fedele, L.; de Gennaro, R.; Langella, A.; de Gennaro, M.; Mercurio, M. Urban geology: Relationships between geological setting and architectural heritage of the Neapolitan area. *J. Virtual Explor.* **2010**, *36*, 26. [CrossRef]
38. Santacroce, R.; Cioni, R.; Marianelli, P.; Sbrana, A.; Sulpizio, R.; Zanchetta, G.; Donahue, D.J.; Joron, J.L. Age and whole rock–glass compositions of proximal pyroclastics from the major explosive eruptions of Somma-Vesuvius: A review as a tool for distal tephrostratigraphy. *J. Volcanol. Geotherm. Res.* **2008**, *177*, 1–18. [CrossRef]
39. Melluso, L.; Scarpati, C.; Zanetti, A.; Sparice, D.; de' Gennaro, R. The petrogenesis of chemically zoned, phonolitic, Plinian and sub-Plinian eruptions of Somma-Vesuvius, Italy: Role of accessory phase removal, independently filled magma reservoirs with time, and transition from slightly to highly silica undersaturated magmatic series in an ultrapotassic stratovolcano. *Lithos* **2022**, *430–431*, 106854. [CrossRef]
40. Rispoli, C.; De Bonis, A.; Esposito, R.; Sossio, F.G.; Langella, A.; Mercurio, M.; Morra, V.; Cappelletti, P. Unveiling the secrets of Roman craftsmanship: Mortars from Piscina Mirabilis (Campi Flegrei, Italy). *Archaeol. Anthr. Sci.* **2020**, *12*, 8. [CrossRef]
41. Dessales, H. *The Villa of Diomedes: The Making of a Roman Villa in Pompeii*; HERMANN: Paris, France, 2020.
42. Dessales, H. Les constructions en tuf jaune dans la Campanie romaine: Usages et modalités d'entretien. In *Sarta Tecta. De l'entretien à la Conservation des Édifices, Antiquité, Moyen Âge, Début de la Période Moderne*; Davoine, C., L'Héritier, M., Péron d'Harcourt, A., Eds.; BiAMA 26; Presses Universitaires de Provence: Aix-en-Provence, France, 2019; pp. 59–69.

heritage

MDPI

Article

The Use of Stone Resources in the Roman Architecture of Oderzo (Treviso, Italy)

Chiara Girotto [1,*], Simone Dilaria [1,2], Caterina Previato [1], Jacopo Bonetto [1] and Claudio Mazzoli [3,*]

1 Department of Cultural Heritage, University of Padova, Piazza Capitaniato 7, 35139 Padova, Italy; simone.dilaria@unipd.it (S.D.); caterina.previato@unipd.it (C.P.); jacopo.bonetto@unipd.it (J.B.)
2 Inter-Departmental Research Centre for the Study of Cement Materials and Hydraulic Binders (CIRCe), Via Gradenigo 6, 35131 Padova, Italy
3 Department of Geosciences, University of Padova, Via Gradenigo 6, 35131 Padova, Italy
* Correspondence: chiara.girotto.3@phd.unipd.it (C.G.); claudio.mazzoli@unipd.it (C.M.)

Abstract: This paper presents the results of the analyses conducted on 46 stone samples collected from Roman buildings in Oderzo, a small town located in the heart of the eastern Venetian plain (29 samples), and from architectural artifacts preserved at the local archeological museum "Eno Bellis" (17 samples). The aim of this study is to identify the types and provenance of the stones used for architectural purposes in Roman times in the city of Oderzo (ancient *Opitergium*). All the materials were petrographically characterized using a multi-analytical approach, including polarized light optical microscopy (PLM). Moreover, volcanic rock samples were analyzed via X-ray fluorescence (XRF) and quantitative phase analysis via X-Ray powder diffraction (QPA-XRPD) to obtain more detailed mineralogical and geochemical characterizations. These methods proved valuable for better determining the provenance of the materials. The results allowed us to determine the quarrying areas that *Opitergium* mostly relied upon in antiquity for sourcing building materials, as well as the stone trade networks in which the city was integrated. Preliminary findings indicate a higher frequency of stones sourced from outcrops along the Prealpine Arc of north-eastern Italy and Istria, including Aurisina limestone (Trieste Karst), and micritic limestones possibly quarried in the Istrian peninsula for architectural artifacts. Conversely, lithotypes from north-western Prealps appear to have been used less frequently. The volcanic rock samples were entirely sourced from various quarry sites in the Euganean Volcanic District in the Veneto region.

Keywords: Oderzo; stone resources; Roman architecture; PML; XRF; Discriminant analysis; QPA-XRPD

1. Introduction

1.1. Objectives of the Research

In the framework of a broader research project on the Roman architecture of *Regio X* (*Venetia et Histria*) conducted over the past few years by the Department of Cultural Heritage in collaboration with the Department of Geosciences of the University of Padova [1–6], this study aims to determine the provenance of the stones used in the city of Oderzo (*Opitergium*) for architectural purposes during the Roman Age and their trade network. To do so, we collected and analyzed 46 stone samples from two of the main archeological areas in the city (the forum and the so called "Ex Carceri" area), as well as from architectural elements preserved in the archeological museum "Eno Bellis" in Oderzo. This study focuses on the use of local stones. The region (Figure 1) is, in fact, naturally rich in resources, including stones, which have been widely exploited since pre-Roman times. These primarily consist of sedimentary rocks, specifically limestones, which outcrop in the Berici Hills, the Venetian Prealps, the Triestine Karst, and the Istrian peninsula [1,7,8]. Additionally, the main outcrop for the sourcing of volcanic rock materials consists of the Oligocene Euganean volcanic district in the Veneto region [9]. Both sedimentary and volcanic rocks were extensively used in Roman times for constructing structures, infrastructure, and artifacts, as well as for general building purposes.

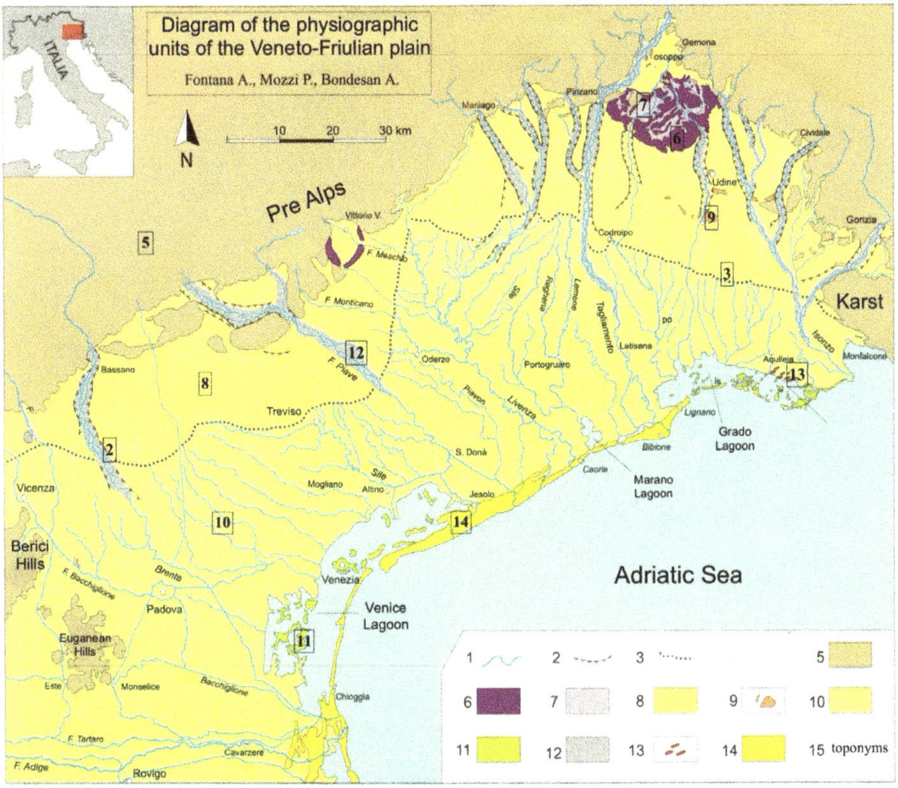

Figure 1. Diagram of the physiographic units of the Veneto–Friulan plain: (1) hydrography; (2) river slopes; (3) springs line; (5) Euganean Hills, Berici Hills, Prealps; (6) Morainic ridges of the Piave and Tagliamento river basins; (7) intermorainic depressions; (8) upper plain; (9) low plain; (10) tectonic terraces; (11) lagoon islands; (12) gravel riverbeds; (13) sand uplands; (14) delta coastal system; (15) toponyms (modified from [10]).

An in-depth scientific study of the quarrying and trading strategies, along with the data derived from it, has considerable informational potential. In fact, identifying the quarrying basins of origin makes it possible to reconstruct the trade routes that shaped the region during Roman times and to identify possible preferential relationships with certain quarries, which could indicate close economic ties between the builders, the quarry workers, and those responsible for transporting the material to the site. Reconstructing such dynamics is crucial as the quarrying phase and, in particular, the transportation of the stone involved considerable costs that undoubtedly weighed heavily on the builders' financial budgets.

1.2. State of the Art

For a long time, research on Roman architecture has focused almost exclusively on the formal, stylistic, and chronological aspects of ancient construction. In many studies conducted between the 19th and 20th centuries, the meticulous analysis of these stylistic aspects was not matched by equal attention to the construction materials and techniques used in ancient buildings, nor to the related dynamics of supply, processing, and trade of natural resources, including stone materials.

The study of stone resources and their use in Roman architecture is a relatively recent topic. In regards to Northern Italy, the first to propose a review of the quarrying basins in the Veneto region, as well as a distribution map of the different lithotypes used in urban centers of the Roman Age, is A. Buonopane [11]. In more recent years, the University of Padua started a research project on the stone resources of the X Regio, investigating the dynamics of their extraction, circulation, and use. The research has, in some cases, focused on specific urban centers and related territories, such as Aquileia [1] and Verona [12], where ancient buildings have been examined through in-depth studies which included archaeometric analysis on stone materials. In addition to these synthesis works, several contributions have focused on the region's stone resources, emphasizing different perspectives, such as the trading and diffusion of stone types [2–4], provenance determination studies [6,13–19], and transport routes and commercial dynamics of stone materials [20,21].

Despite such remarkable progress, the implementation of an interdisciplinary approach—specifically, a multi-analytical protocol that includes petrographic and archaeometric analyses—remains rare in the study of stone materials used in the Roman architecture of the region.

2. History and Architecture of Roman Oderzo

Located in the heart of the Eastern Venetian plain, Oderzo stands on a hillock to the right of a wide bend of the Monticano river. The area is bordered to the west by the Piave river and to the east by the Livenza river [22,23]. The site's advantageous location, relatively close to the coastline and accessible in ancient times via the Navisego Vecchio-Piavon and Monticano rivers [24], is undoubtedly one of the reasons for the settlement's longevity. Its development spans over three millennia, beginning between the 10th and 9th centuries BC, where it served as an important crossroad and communication hub connecting eastern Veneto, the Alpine region, and the Euganean area. During this period, the settlement occupied the top of the hillock and consisted of a cluster of huts. The crossroad nature of the site facilitated the early development of a complex road network, whose main routes continued to be used even in Roman times [25].

The adaptation of the settlement layout to Roman urban standards occurred relatively quickly (Figure 2). By the late 2nd century to early 1st century BC, the city was already provided with a forum square. This period of architectural expansion was further enhanced by the nearby *Via Postumia*, constructed in 148 BC by the consul Spurius Postumius Albinus [26].

In the second half of the 1st century BC, when the region was granted Roman citizenship, the newly established *municipium Opitergium* underwent an extensive urban restructuring plan that continued throughout the 1st century AD. This project led to the construction of impressive public architectural complexes, including a monumental square (*forum*) [27], a bath building [28], a temple [29], and many prestigious houses (*domus*) [25,30,31]. A dense network of infrastructure, including paved roads [32], a river dock [33], and city walls [34], also date to this period. During this time, as this study shows, the import of stone material increased. The stone used in this phase of reconstruction and renovation came from quarries throughout the region, including those further away from the city.

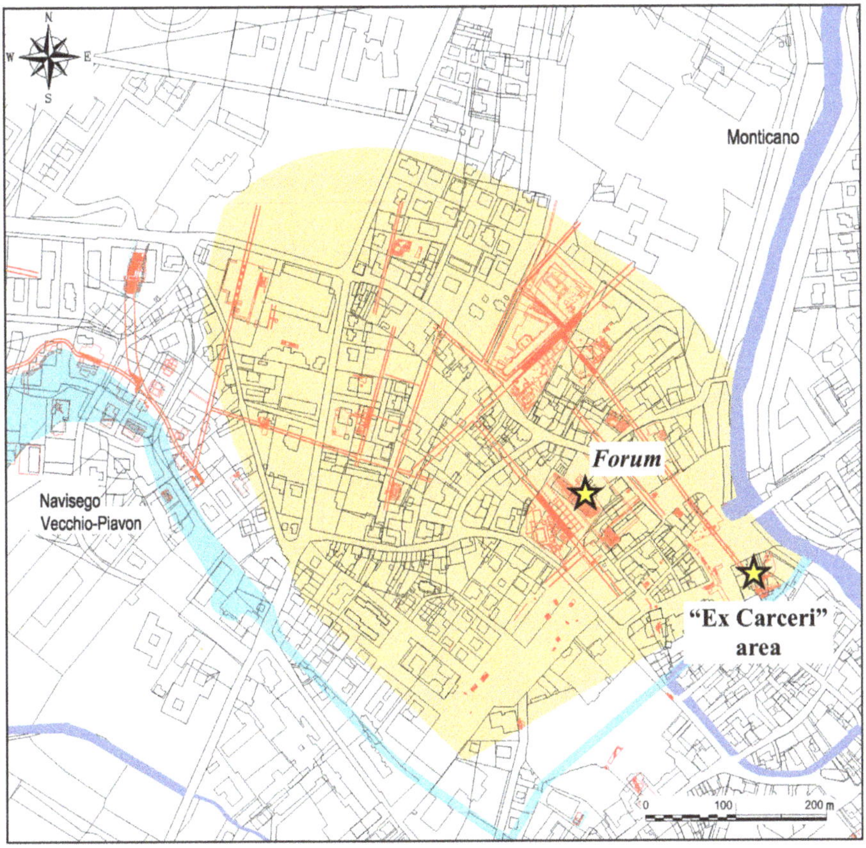

Figure 2. Map of Roman Opitergium, with an indication of the archeological areas analyzed in this study [modified from [25]] (blue = rivers; yellow = ancient city area; red = ancient structures and infrastructure).

The settlement's development eventually slowed down due to the looming barbarian invasion, which threatened the city as early as the 2nd century AD, notably with the incursions of the Quadi and Marcomanni. This also led to a decrease in the import of stone material, as evidenced by the rare stone samples taken from buildings dated to this period and analyzed in this research.

A mosaic floor depicting hunting scenes [35], dated to the 3rd century AD, indicates a brief revival of residential construction during the Late Antique period just before a decline in the following centuries. At this stage, there was a modest increase in stone imports, particularly from the nearby Prealps.

Our knowledge of the Late Antique and Medieval phases of the city comes from the archeological evidence found in the so-called Ex Carceri area [36]. In the 7th century AD, a new defense wall was constructed using the remnants of Roman structures (including a section of the Augustan walls, a paved road, and a city gate), reusing numerous stone elements from Roman funerary contexts, such as cylindrical altars, decorated stelae, and an inscribed statue base [37]. The need for robust fortifications came from the Lombard invasion that led to the destruction of the city in 667, as recounted by Paolo Diacono in the *Historia Langobardorum* (IV, 38, 45).

Despite the city's destruction, occupation of the site continued, and new restructuring projects soon began. A keep, built on the southwestern corner of the 7th century AD wall using large blocks of Euganean trachyte and limestone taken from the ruins of monumental buildings of Roman *Opitergium*, was realized between the 8th and 9th centuries [38]. These structures were later utilized as foundations for new buildings and infrastructure, including a tower house, a new city wall, and ultimately, the Medieval castle. The prison tower, which lends its name to the area, was erected in 1797 on the castle's ruins.

3. Materials and Methods

3.1. Sampling

For the purpose of this research, 46 stone samples were taken, 29 from structures preserved in the archeological areas of the forum and the Ex Carceri and 17 from stone artifacts preserved in the Archeological Museum "Eno Bellis" in Oderzo (Figure 3). In particular, 16 pavement elements, with 8 from the *cavaedium* of the city gate (OD_01; OD_02; OD_03; OD_04; OD_05; OD_06; OD_09; OD_10), 1 from a paved road (OD_11), 1 from the pavement of the basilica (OD_19), 6 from the forum square (OD_25; OD_27; OD_28; OD_31; OD_33; OD_35); 2 small pillars (OD_13; OD_14); 1 drainage lid (OD_29); 1 drainage filter (OD_18); 3 thresholds, with 2 from the *cavaedium* of the city gate (OD_07; OD_12) and 1 from the basilica (OD_20); and 23 architectural elements, with 1 acroterium (OD_52), 2 architraves (OD_45; OD_51), 5 ashlars (OD_08; OD_15; OD_16; OD_17; OD_36), 4 capitals (OD_43; OD_49; OD_54; OD_55), 1 column base (OD_37), 3 columns (OD_38; OD_41; OD_50), 5 cornices (OD_22; OD_23; OD_44; OD_46; OD_53), 1 frieze fragment (OD_39), and 1 fragment of semi-column (OD_42). The operation was carried out via mechanical removal, using a chisel, of centimeter-sized fragments of stone material. Where possible, elements characterized with pre-existing gaps were favored, which allowed for the easy identification and removal of the material, as well as minimal impact on structures. All samples came from structures and artifacts dated to Roman times [39]. In buildings with more than one construction phase, when possible, samples were taken from structures of all the different phases in order to identify elements of either continuity or discontinuity in the selection and use of stone materials.

3.2. Analythical Method

To gather as much information as possible from the stone samples, we applied an analytical protocol that was already established [40]; it involved characterizing different stone types using various archaeometric approaches, including textural and microfaunal analyses through optical microscopy and geochemical analyses. The results are then compared with established data on the petrographic and geochemical signatures of quarry sites to determine the likely provenance of the rock samples.

The preliminary macroscopic identification allowed us to classify the samples into two groups: one consisting of 30 sedimentary rock samples and the other of 16 volcanic rock samples, predominantly Euganean trachyte. Subsequently, the samples underwent two different characterization protocols. Both sedimentary rocks, primarily carbonate, and volcanic rock samples were first analyzed petrographically using PLM to identify their petrofacies, which are useful for determining their geological formation.

Moreover, a more detailed analytical approach was adopted for volcanic rocks, which allowed us to identify the specific quarry site for most of the samples.

Figure 3. Selection of archeological structures and artifacts sampled in the archeological areas and in the archeological museum "Eno Bellis" in Oderzo: (**a**) OD_19 = drainage lid; (**b**) OD_55 = capital; (**c**) OD_39 = decorated architrave; (**d**) OD_25 = paving slab; (**e**) OD_41 = column.

The volcanic rock samples were therefore characterized from a geochemical point of view using XRF (X-ray fluorescence) analysis. The quantitative data of major and trace elements thus obtained were then compared with databases of geological samples taken from known quarries in the Euganean area [41] to assess correlations with the quarry basins. Both the geological quarry samples and archeological ones were analyzed via XRF by adopting the same analytical protocol as reported in [42]. Reference samples constitute the geochemical databases implemented by the Department of Cultural Heritage and the Department of Geosciences of the University of Padua, comprising modern geological samples directly collected from Euganean quarries coupled with archeological samples (analyzed in [40] and previous studies) unequivocally attributed to quarry outcrops. The samples were categorized in groups according to their chemical composition via cluster analyses and then compared in bivariate scatterplots to reference samples in order to better identify their geological provenance from Euganean volcanic outcrops using a selection of discriminant chemical elements selected among major elements, specifically REE and HFSE. For further confirmation, the data obtained were analyzed using discriminant analysis, with the same pattern of major and trace elements used to produce the scatterplots and PCA/cluster analyses as the discriminant variables. The archeological samples were again compared with 218 samples of known origin. Discriminant analysis (DA), using the tools available in Statgraphics Centurion Pro 19, enabled the probabilistic identification of quarry sites by providing percentage-based likelihoods of origin. When the probability for a primary site did not reach 100% (unequivocal attribution), a secondary site was also suggested on a probabilistic basis. This approach, frequently used by our team for volcanic rock identification [6,40], provides robust probabilistic provenance attribution based on geochemical data. Finally, a coherent selection of samples was subjected to QPA-XRPD (X-ray powder diffraction) analysis to observe possible discrimination markers based on the bulk mineralogy of the samples.

3.3. Analytical Equipment

All analyses were conducted using equipment located at the Department of Geosciences, University of Padova.

The PLM-TL (transmitted light) analyses were performed on rock samples prepared on 30 μm thin sections using a Leica DM750 P microscope equipped with an integrated FLEXACAM I5 digital camera (Leica Geosystems, Heerbrugg, Switzerland).

The XRF analyses were performed on volcanic rock samples with a WDS Panalytical Zetium sequential spectrometer (Malvern Panalytical, Malvern, UK), operating under vacuum conditions and equipped with a 2.4 kW Rh tube. The samples were calcined to determine their loss on ignition (L.O.I.) by placing them in a muffle furnace at 860 °C for about 20 min and then at 980 °C for about 2 h. The samples for the actual XRF analyses were then prepared in beads using a lithium tetraborate ($Li_2B_4O_7$) flux with a dilution of 1:10, and melting was carried out using a Claisse Eagon 2 bead mill (maximum temperature achieved of about 1150 °C). The calculated major elements are Si, Ti, Al, Fe, Mn, Mg, Ca, Na, K, and P (expressed as percentages of the relative oxides). The L.O.I. was calculated separately. The calculated trace elements (expressed in ppm) are Sc, V, Cr, Co, Ni, Cu, Zn, Ga, Rb, Sr, Y, Zr, Nb, Ba, La, Ce, Nd, Pb, Th, and U. Instrumental precision (defined by repeated analyses on the same sample) is within 0.6% relative for major elements and within 3.0% relative for trace elements. Detection limits for Al, Mg, and Na are within 0.01%, within 0.2% for Si, and within 0.005% for Ti, Fe, Mn, Ca, K and P; for trace elements, they are (in ppm) Sc = 3, V = 5, Cr = 6, Co = 3, Ni = 3, Cu = 3, Zn = 3, Ga = 3, Rb = 3, Sr = 3, Y = 3, Zr = 3, Nb = 3, Ba = 10, La = 10, Ce = 10, Nd = 10, Pb = 5, Th = 3, and U = 3.

The QPA-XRPD analyses, performed on a selection of representative volcanic rocks, were performed using a Bragg–Brentano θ-θ diffractometer PANalytical X'Pert PRO (Malvern Panalytical, Malvern, UK), equipped with a real-time multiple strip (RTMS) detector (Malvern, UK, PANalytical, Malvern, UKPIXcel by Panalytical). The quantification of both crystalline and amorphous content was obtained through the addition of 20 wt% of zincite to the powders as an internal standard. Then, a quantitative phase analysis (QPA) was performed using the Rietveld method [43]. Refinements were carried out with the TOPAS software (version 4.1) by Bruker AXS (Bruker Corporation, USA). The observed Bragg peaks in the powder patterns were modeled through a pseudo-Voigt function, fitting the background with a 12-coefficient Chebyshev polynomial. For each mineral phase, lattice parameters, Lorentzian crystal sizes, and scale factors were refined. Any residual preferred orientation effect was modeled during the refinement with the March Dollase algorithm [44]. The starting structural models for the refinements were taken from the International Crystal Structure Database (ICSD). For the quantification of low-angle smectite-like peaks, detected in some samples, the BGMN-Profex software, version 5.1.8 (Brandt, Germany) with dedicated fitting functions for Rietveld refinements of clay phases was adopted [45].

4. Results

4.1. PLM (Polarized Light Optical Microscopy)

4.1.1. Sedimentary Rocks

The analytical approach adopted for the study of sedimentary rocks involves the analysis of samples using PLM in order to obtain a petrographic profile to identify, by comparison with databases of samples of known provenance, the extraction basin of origin. The samples are mostly pertinent to carbonate rocks, among which are Aurisina limestone (nine samples: OD_12; OD_20; OD_23; OD_39; OD_44; OD_45; OD_51; OD_52; OD_53), Scaglia Rossa (six samples: OD_06; OD_08; OD_10; OD_16; OD_17; OD_19), and Vicenza stone (three samples: OD_41; OD_54; OD_55). For an additional eight samples, the identification with a specific lithotype and corresponding source basin is uncertain; a total of five of these samples (OD_37; OD_38; OD_42; OD_43; OD_50) exhibit characteristics consistent with Karst limestones, while the remaining three samples (OD_15; OD_36; OD_49) resemble compact limestones quarried from the Istrian peninsula. Among sedimentary rocks, there are four sandstone samples (OD_18; OD_22; OD_35; OD_46).

Aurisina Limestone

The samples analyzed in this study (OD_12; OD_20; OD_23; OD_39; OD_44; OD_45; OD_51; OD_52; OD_53) are predominantly classified as grainstone, with abundant fragments of rudists and a smaller percentage of benthic foraminifera (including Miliolida and Textulariida). Fragments of green algae (*Thaumatoporella parvovesiculifera*), echinoderms, peloids, and rare intraclasts are also present (Figure 4a). They are compatible with the quarry basin located in the Municipality of Duino-Aurisina, which is the largest and most heavily exploited in the Trieste Karst. The lithological unit quarried in the area dates to the transition from the Middle to the Upper Turonian (Late Cretaceous) and is characterized by compact limestones that are very rich in fragments of *Radiolites*, Neitheidae, and Chondrodontidae with predominantly grayish-light hues [46,47]. The lithological unit includes different varieties of limestones, which are commonly known by their trade names: "Aurisina Chiara", "Aurisina Fiorita", "Aurisina Granitello", and "Roman Stone". These varieties are distinguished based on texture, grain size, and bioclastic components [48,49].

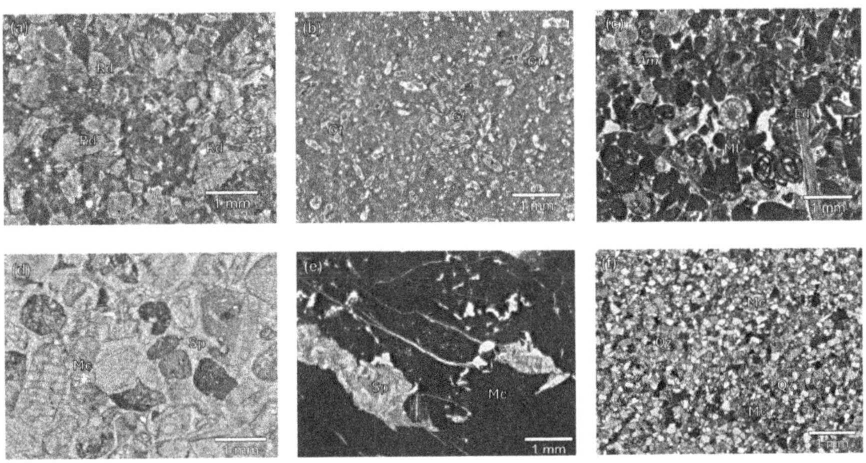

Figure 4. PLM micrographs (plane polarized light) of a selection of samples of sedimentary rocks. (**a**) OD_12; (**b**) OD_08; (**c**) OD_41; (**d**) OD_37; (**e**) OD_36; (**f**) OD_18; Rd, Rudists; *Gt, Globotruncana; Gl, Globigerina* sp.; *Am Amphistegina* sp.; Ml, Miliolida; Ed, Echinoderms, Mc, micritic carbonates; Sp, Sparite; Qz, Quartz.

Scaglia Rossa

The samples (OD_06; OD_08; OD_10; OD_16; OD_17; OD_19) analyzed mainly fall under the definition of wackestone, with abundant planktonic foraminifera (Globotruncanidae, Globigerinidae, and Heterohelicidae), less frequent calcispheres, and a rare presence of iron hydroxides and fissure-filling clay minerals, compatible with the Scaglia Rossa formation.

In the Veneto region, the Scaglia Rossa formation outcrops are found in the area of the Euganean [10] and Lessini Hills [50], as well as in the Possagno/Valdobbiadene area [51]. This formation is generally classified as a wackestone, characterized by the abundant presence of planktonic foraminifera and micritic cement (Figure 4b). In smaller quantities, calcispheres, radiolarians, and dinoflagellates are noted, while calpionellids, echinoderms, and bivalves are rarer.

Vicenza Stone

Three of the samples analyzed are Vicenza stone (OD_41; OD_54; OD_55), a soft limestone that outcrops mainly in the Berici Hills and the Lessini Mountains [9,52]. All three samples have a petrographic profile that is consistent with the Oligocene formation of Castelgomberto and are partially comparable to *Petrofacies 1* that was identified in the Roman site of Adria [40]. Specifically, these samples are red algae-encrusted grainstones, with an abundant presence of benthic foraminifera (including Miliolida), Rotaliida (*Amphistegina* sp., *Asterigerina* sp.), and, in lower concentrations, hyaline foraminifera. Some echinoderms, bivalves, and bryozoans are also noted. The samples are characterized by high porosity (Figure 4c).

Trieste Karst (?)

For five samples (OD_37; OD_38; OD_42; OD_43; OD_50), the basin of origin could not be identified with certainty. The petrographic data do not allow for precise attribution to a single quarry site. In fact, the lack of definitive elements and index fossils for lithotype recognition complicates the attempt to trace these samples to a specific outcrop. However, their petrographic characteristics show some consistency with the lithotypes quarried in the Trieste Karst.

The samples are mostly grainstones with abundant well-rounded clasts, which are sometimes recrystallized and surrounded by a micritic rim of microbial origin. Intraclasts of crystalline limestones are also present in almost every sample, as well as in well-developed sparitic cement (Figure 4d).

In two samples (OD_42; OD_43), traces of organogenic detritus can be recognized, including fragments of rudists, echinoderms, and algae (*Bacinella irregularis*).

Istrian Stone (?)

For three other samples (OD_15; OD_36; OD_49), the basin of provenance is also uncertain. The three samples are mudstones; they are characterized by the presence of fenestrae filled with sparitic calcite and rare peloids. Rare fossil fragments are also present, but they are not easily recognizable (Figure 4e). Only one sample (OD_15) possibly contains sponge spicules. The petrographic profile is partly consistent with that of compact limestones quarried in the Istrian peninsula [9].

Sandstone

Among the sedimentary rock samples, four are sandstones (OD_18; OD_22; OD_35; OD_46). They are medium-grained polygenic litharenites, predominantly consisting of carbonate clasts (80–90%) and a subordinate siliciclastic fraction (10–20%) made up of quartz and flint clasts, a common feature across all the samples. The clasts show a good degree of compaction and are well cemented, resulting in generally low porosity (Figure 4f).

The closest rock outcrops compatible with the archeological samples are those of the Venetian Prealps, where the Miocene formation of Molassa Sudalpina [51] is exposed.

4.1.2. Volcanic Rocks

All volcanic rock samples are petrographically consistent with the volcanic rocks of the Oligocene Euganean volcanic district in the Veneto region. The results of this study grouped the samples into four macrogroups based on specific textural and petro-mineralogical features.

Group 1 includes nine samples (OD_03; OD_04; OD_07; OD_11; OD_25; OD_27; OD_28; OD_31; OD_33) (Figure 5a). These are trachytes with porphyritic/glomeroporphyritic textures and a microlithic matrices. The main phenocrysts are anorthoclase, plagioclase, and biotite. Their petrographic profile corresponds well to the outcrops of M. Oliveto/ M. Alto.

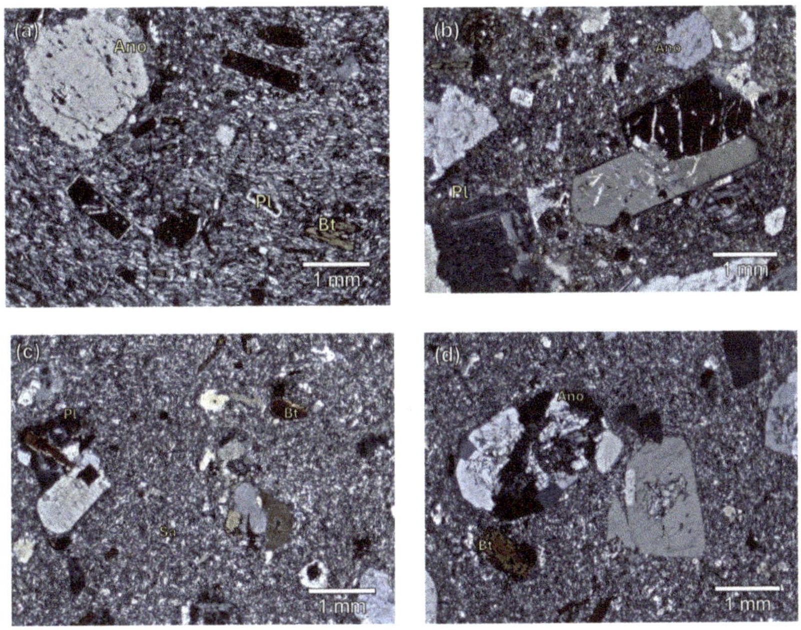

Figure 5. PLM micrographs (crossed polarized light) of a selection of samples of volcanic rocks. (**a**) OD_01; (**b**) OD_03; (**c**) OD_05; (**d**) OD_13; Ano, Anorthoclase; Bt, Biotite; Pl, Plagioclase; Sa, Sanidine.

Group 2 comprises four samples (OD_01; OD_02; OD_09; OD_29), whose petrographic and textural features are consistent with rocks from the Monselice quarry (Figure 5b). These trachytic rocks are characterized by phenocrysts of anorthoclase, plagioclase, and, to a lesser extent, sanidine. Accessory minerals include apatite, zircon, and opaque minerals.

Group 3 consists of a single sample (OD_05, Figure 5c), whose petrofabric does not allow for clear attribution to a specific outcrop within the Euganean district. This trachyte features a fine-grained matrix, with sanidine phenocrysts predominating over plagioclases and anorthite, as well as biotite phenocrysts among the accessory phases. The last group (Group 4) includes two samples (OD_13 and OD_14), characterized by rocks with porphyritic structures and microcrystalline matrices. These rocks contain large-grained quartz crystals embedded within the groundmass (Figure 5d).

4.2. Geochemical and Mineralogical Analyses of Volcanic Rock Samples

To achieve a more precise and detailed assessment of the provenance of the Euganean volcanic rock samples from supply basins, all analyzed samples of these lithotypes underwent chemical and mineralogical investigations via XRF and QPA-XRPD. These analyses enabled a more accurate lithotype classification and supported the attribution of samples to known quarries by assessing their geochemical and mineralogical compatibility with geological reference samples (markers) from the database of the Euganean volcanic rocks implemented by the Department of Geosciences and the Department of Cultural Heritage of the University of Padova.

According to the TAS diagram (Figure 6), most samples display a trachytic geochemical profile, while only two samples (OD_13 and OD_14) show a rhyolitic profile, which is due to their high SiO_2 content (nearly 70% wt.), as documented through the XRF analysis (Supplementary Table S1).

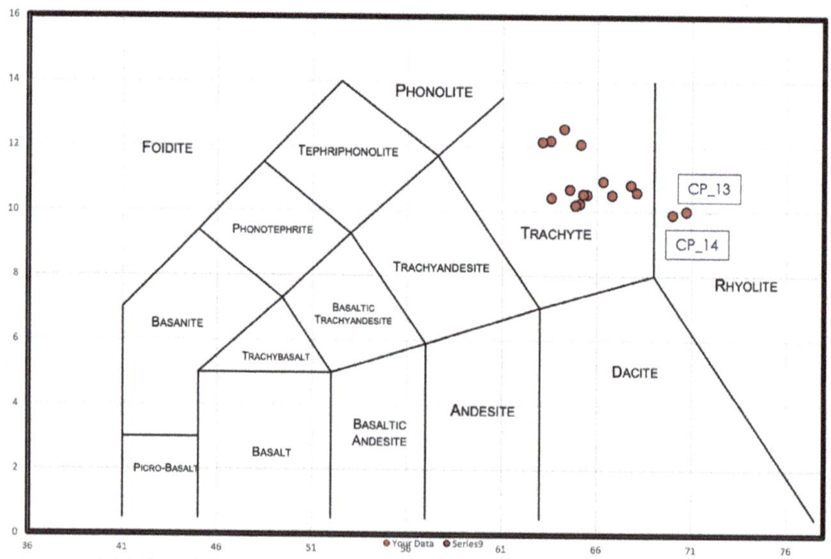

Figure 6. Total alkali vs. silica (TAS) geo-chemical distribution diagram of the volcanic rock samples analyzed using XRF (after [53]). The samples within the rhyolite field are labeled.

Cluster analysis, performed using the furthest neighbor method with Euclidean distance, was applied to these samples based on descriptive geochemical values determined using XRF, which allowed for the geochemical validation of the groupings determined after petrographic observations (Figure 7). Group 2 is notably composed of samples OD_01, 09, 29, and 02, while Group 1 includes samples OD_03, 28, 07, 27, 04, 11, 31, and 33, with sample OD_25 included in this cluster but at a greater distance (>4) from the other ones. This analysis also clearly identifies the two rhyolitic samples, OD_13 and OD_14 (Group 4), as well as the more isolated sample CP_05 (quartz trachyte) with a distance > 3 (Group 3).

Precise provenance from quarry sites within the Euganean district was further determined by comparing the geochemical profiles of the samples from Oderzo with that of the reference ones. This was first determined by using selected bivariate scatterplots reporting selected chemical elements useful for differentiating the Euganean trachytes and rhyolites. The discriminant variables considered for the scatterplots were selected following the methodology reported by Dilaria [40] and by building on the approaches of Germinario et al. [54–56]. For instance, the Nb vs. V plot (Figure 8a) effectively distinguishes Group 2 samples as compatible with M. Merlo trachyte; they are clearly identifiable because of the high Nb concentrations (>120 ppm). The K_2O vs. TiO_2 plot (Figure 8b) is useful for isolating the trachyte from M. Oliveto, attributed to low TiO_2 and K_2O content. This encompasses Group 1 samples, including OD_25. Despite being somewhat separated both in the scatterplot as well as from the cluster analysis, OD_25 completely aligns with reference samples from M. Oliveto. This confirms the compatibility of this sample, along with the others of Group 1, with this outcrop rather than the nearby M. Alto, which had initially been suggested via petrographic analysis as a possible source for the rocks in this group. In the Zr vs. Rb scatterplot (Figure 8c), the clustering for M. San Daniele samples is consistent within Zr intervals around 450–500 ppm, with sample OD_05 falling within its range. Further compatibility is coherent in the K_2O vs. TiO_2 plot. For the rhyolitic samples OD_13 and OD_14, the Zr vs. Rb plot showing very low Zr content, below 300–350 ppm, and high Sr values, over 170 ppm. They clearly overlap with reference samples from Euganean rhyolites, though without a clear association with a specific outcrop.

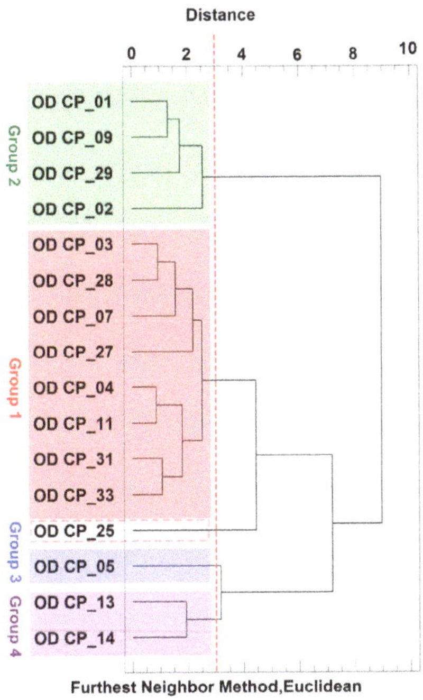

Figure 7. Results of the cluster analysis (furthest neighbor method, Euclidean) of the Euganean volcanic rock samples from Oderzo, based on XRF data.

Discriminant analysis (DA) further validated the sample associations with quarry districts (Table 1, see coefficient factors in Supplementary Table S2). A series of discriminant chemical elements was used as independent variables, with samples already associated with quarry sites (samples in the reference database) as the classification factors. The selected elements for distinguishing samples are Na_2O, K_2O, TiO_2, V, Rb, Sr, Zr, Nb, Al_2O_3, Nd, Ce, which are useful for trachyte discrimination, and SiO_2. Ba is indicative of rhyolite differentiation. The DA achieved a 94% accuracy rate in correctly classifying 218 reference samples, highlighting the effectiveness of these variables in distinguishing the geochemical profiles of each quarry site. The results of DA, presented in Table 2, confirm the scatterplot observations and identify the best-matching quarry sites for previously unassigned archeological samples. In detail, the provenance of volcanic rock samples from Groups 1 and 2 was further confirmed through DA. Moreover, DA proved the attribution of OD_05 (Group 3) to M. San Daniele with a 100% confidence rate, while the two rhyolitic samples OD_13 and OD_14 (Group 4) were attributed to the Forche del Diavolo rhyolite outcrop in the Western part of the Euganean district with confidence levels of 99.9% and 100%, respectively. The only exception was sample OD_07, which was first associated via DA with M. Alto, though M. Oliveto is reported as its second match in probabilistic terms. Moreover, M. Alto often appeared as the second most probable source for M. Oliveto samples, underscoring the challenge of differentiating these two outcrops based solely on their geochemical footprint.

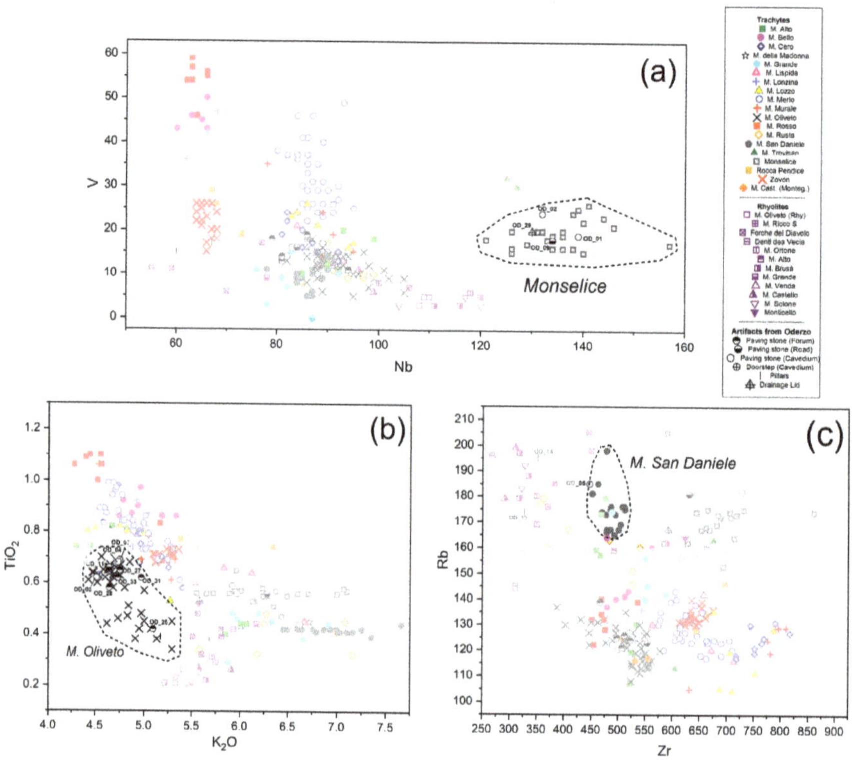

Figure 8. (**a**) V vs. Nb, (**b**) TiO_2 vs. K_2O, and (**c**) Rb/Zr scatterplots used for the provenance discrimination of the archeological volcanic rock samples from Oderzo. Reference samples constitute the geochemical databases implemented by the Department of Cultural Heritage and the Department of Geosciences of the University of Padua, comprising modern geological samples directly collected from Euganean quarries coupled with archeological samples (analyzed in [40] and previous studies) unequivocally attributed to quarry outcrops. Indicative distributions of samples from selected quarry sites are highlighted by dotted lines. Certain archeological samples from Oderzo labeled in the scatterplots are fully described in the body of the text.

Table 1. Results of the discriminant analysis of the archeological volcanic stone samples from Oderzo (see Supplementary Table S2 for analytical details).

Sample	Provenance (1st)	Prob. (%)	Provenance (2nd)	Prob. (%)
OD_01	Monselice	100		null.
OD_02	Monselice	100		null.
OD_03	M. Oliveto	99.72	M. Alto	0.27
OD_04	M. Oliveto	90.31	M. Alto	9.52
OD_05	M. San Daniele	100		null.

Table 1. *Cont.*

Sample	Provenance (1st)	Prob. (%)	Provenance (2nd)	Prob. (%)
OD_07	M. Alto	88.58	M. Oliveto	10.65
OD_09	Monselice	100		null.
OD_11	M. Oliveto	97.32	M. Alto	1.37
OD_13	Forche del Diavolo	99.91	Denti dea Vecia	0.09
OD_14	Forche del Diavolo	100		null.
OD_25	M. Oliveto	94.32	M. Alto	5.68
OD_27	M. Oliveto	100		null.
OD_28	M. Oliveto	99.84	M. Alto	0.15
OD_29	Monselice	100		null.
OD_31	M. Oliveto	98.27	M. Alto	1.72
OD_33	M. Oliveto	99.68	M. Alto	0.22

Table 2. List of the analyzed samples and their provenance according to the archaeometric analyses (Latin words appear in italics).

Sample	Artifact	Structure	Age	Rock Type	Extraction Basin/Quarry Site
OD_01	Paving slab	*Cavaedium* urban gate	1st c. BC–1st c. AD	Euganean trachyte	Euganean Hills, Monselice
OD_02	Paving stone	*Cavaedium* urban gate	1st c. BC–1st c. AD	Euganean trachyte	Euganean Hills, Monselice
OD_03	Paving stone	*Cavaedium* urban gate	1st c. BC–1st c. AD	Euganean trachyte	Euganean Hills, M. Oliveto
OD_04	Paving stone	*Cavaedium* urban gate	1st c. BC–1st c. AD	Euganean trachyte	Euganean Hills, M. Oliveto
OD_05	Paving stone	*Cavaedium* urban gate	1st c. BC–1st c. AD	Euganean trachyte	Euganean Hills, M. San Daniele
OD_06	Paving stone	*Cavaedium* urban gate	3rd–4th c. AD	Scaglia Rossa	Prealps?
OD_07	Threshold	*Cavaedium* urban gate	3rd–4th c. AD	Euganean trachyte	Euganean Hills, M. Oliveto
OD_08	Architectural element (ashlar block)	*Cavaedium* urban gate	3rd–4th c. AD	Scaglia Rossa	Prealps?
OD_09	Paving stone	*Cavaedium* urban gate	3rd–4th c. AD	Euganean trachyte	Euganean Hills, Monselice
OD_10	Paving stone	*Cavaedium* urban gate	3rd–4th c. AD	Scaglia Rossa	Prealps?
OD_11	Paving stone	Road	3rd–4th c. AD	Euganean trachyte	Euganean Hills, M. Oliveto
OD_12	Threshold	*Cavaedium* urban gate	3rd–4th c. AD	Aurisina limestone	Trieste Karst
OD_13	Architectural element (pillar)	Roadside	3rd–4th c. AD	Rhyolite	Euganean Hills, Forche del Diavolo

Table 2. *Cont.*

Sample	Artifact	Structure	Age	Rock Type	Extraction Basin/Quarry Site
OD_14	Architectural element (pillar)	Roadside	3rd–4th c. AD	Rhyolite	Euganean Hills, Forche del Diavolo
OD _15	Architectural element (ashlar block)	Bizantyne urban wall	7th c. AD	Istria stone?	Istrian Peninsula?
OD_16	Architectural element (ashlar block)	Bizantyne urban wall	7th c. AD	Scaglia Rossa	Prealps?
OD_17	Threshold	Bizantyne urban wall	7th c. AD	Scaglia Rossa	Prealps?
OD_18	Water infrastructure (drainage filter)	Drainage channel	1st c. BC–1st c. AD	Sandstone	n.d.
OD_19	Paving stone	*Basilica Forum*	1st c. BC–1st c. AD	Scaglia Rossa	Prealps?
OD_20	Paving stone	*Basilica Forum*	1st c. BC–1st c. AD	Aurisina limestone	Trieste Karst
OD_22	Architectural element (cornice)	*Basilica Forum*	1st c. BC–1st c. AD	Sandstone	n.d.
OD_23	Architectural element (cornice)	*Basilica Forum*	1st c. BC–1st c. AD	Aurisina limestone	Trieste Karst
OD_25	Paving stone	*Forum*	1st c. BC–1st c. AD	Euganean trachyte	Euganean Hills, M. Oliveto
OD_27	Paving stone	*Forum*	1st c. BC–1st c. AD	Euganean trachyte	Euganean Hills, M. Oliveto
OD_28	Paving stone	*Forum*	1st c. BC–1st c. AD	Euganean trachyte	Euganean Hills, M. Oliveto
OD_29	Water infrastructure (drainage lid)	Drainage channel	1st c. BC–1st c. AD	Euganean trachyte	Euganean Hills, M. Monselice
OD_31	Paving stone	*Forum*	1st c. BC–1st c. AD	Euganean trachyte	Euganean Hills, M. Oliveto
OD_33	Paving stone	*Forum*	1st c. BC–1st c. AD	Euganean trachyte	Euganean Hills, M. Oliveto
OD_35	Paving stone	*Forum*	1st c. BC–1st c. AD	Sandstone	n.d.
OD_36	Architectural element (ashlar block)	Oderzo	Roman Age	Istria stone?	Istrian Peninsula?
OD_37	Architectural element (column base)	Oderzo	Roman Age	Aurisina limestone	Trieste Karst
OD_38	Architectural element (column)	Oderzo	Roman Age	Aurisina limestone	Trieste Karst
OD_39	Architectural element (frieze)	Oderzo	1st c. AD	Aurisina limestone	Trieste Karst

Table 2. *Cont.*

Sample	Artifact	Structure	Age	Rock Type	Extraction Basin/Quarry Site
OD_41	Architectural element (column)	Oderzo	Roman Age	Vicenza stone	Berici Hills
OD_42	Architectural element (semi-column)	Oderzo	Roman Age	Karst limestone?	Trieste Karst?
OD_43	Architectural element (capital)	Oderzo	1st c. AD	Karst limestone?	Trieste Karst?
OD_44	Architectural element (cornice)	Oderzo	Roman Age	Aurisina limestone	Trieste Karst
OD_45	Architectural element (architrave)	Oderzo	Roman Age	Aurisina limestone	Trieste Karst
OD_46	Architectural element (cornice)	Oderzo	Roman Age	Sandstone	n.d.
OD_49	Architectural element (capital)	Oderzo	Roman Age	Istria stone?	Istrian Peninsula?
OD_50	Architectural element (capital)	Oderzo	Roman Age	Aurisina limestone	Trieste Karst
OD_51	Architectural element (architrave)	Oderzo	Roman Age	Aurisina limestone	Trieste Karst
OD_52	Architectural element (*acroterion*)	Oderzo	Roman Age	Aurisina limestone	Trieste Karst
OD_53	Architectural element (cornice)	Oderzo	Roman Age	Aurisina limestone	Trieste Karst
OD_54	Architectural element (column)	Oderzo	Roman Age	Vicenza stone	Berici Hills
OD_55	Architectural element (capital)	Oderzo	Roman Age	Vicenza stone	Berici Hills

The QPA-XRPD analysis (Figure 9) of a selection of representative samples from each identified group highlights certain potential discriminant features among the Euganean volcanic rock samples based on their mineralogy. The presence of newly formed calcite as a secondary alteration in humid environments was considered, as well as that of halite produced as a consequence of salt deposition on the surface of certain samples. These phases were removed from the original mineral phases, and their quantification was recalculated at 100% wt. (Supplementary Table S3). In Figure 9, two samples from Group 1, likely attributable to M. Oliveto, exhibit a similar mineralogical profile. These rocks have a high feldspar content, predominantly plagioclase, with subordinate concentrations of sanidine and anorthoclase, while their SiO_2 content is evidenced by the reduced presence of quartz and cristobalite, a high-temperature polymorph of SiO_2 which can develop in volcanic rocks at temperatures typically above 1470 °C. The cryptocrystalline component of the groundmass in these samples is indicated by the significant amount of the amorphous phase (9.6–12.8 wt%), while the presence of a broad low-angle band of smectite (4.6–7.9 wt%) at around 6.5–7.5 2θ suggests considerable clay alteration processes.

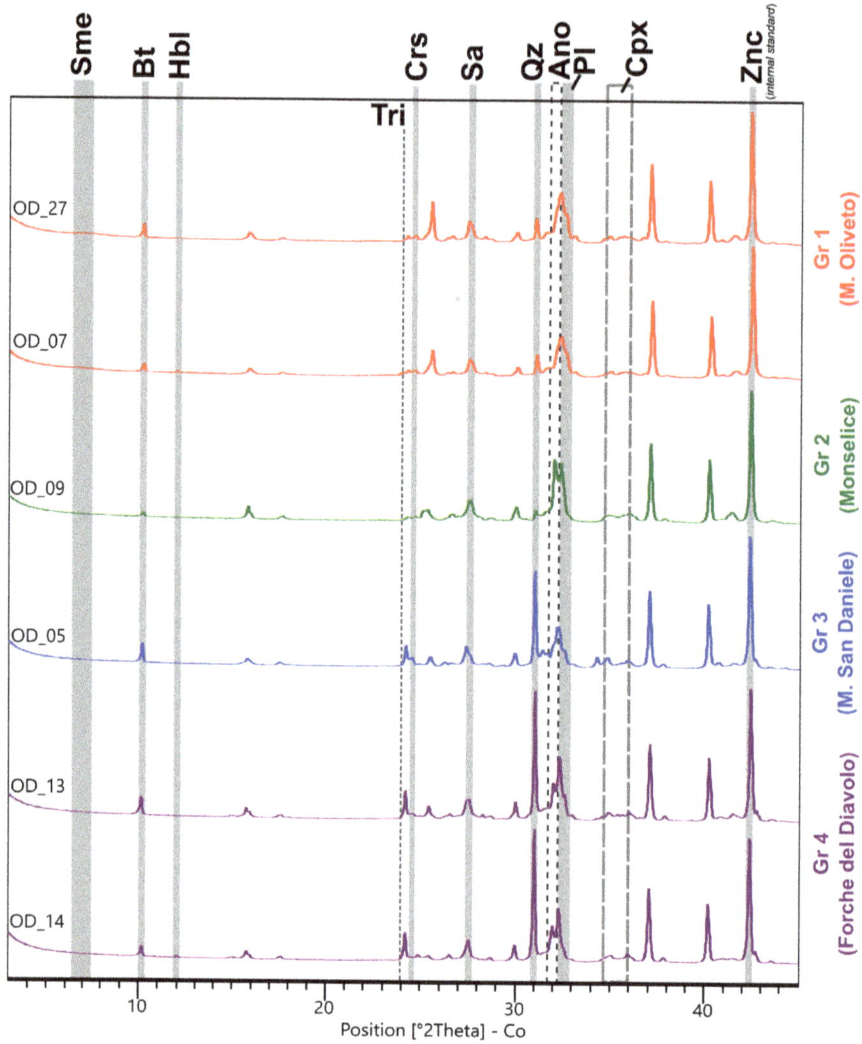

Figure 9. XRPD profiles of selected archeological volcanic rock samples from Oderzo. For each phase, the main peaks are highlighted. The mineral phases are labeled according to [56]: Ano = anorthoclase; Bt = biotite; Cpx = clinopyroxene; Crs = cristobalite; Hbl = hornblende (kaersutite); Ilm = ilmenite; Pl = plagioclase; Qz = quartz; Sa = sanidine; Sme = smectite; Tri = tridymite. Not displayed in the graph are the main peaks of the opaque mineral Ti-magnetite (53.6 2θ), excluding those overlayed by the peaks of the internal standard (Zincite).

Sample OD_09, from Monselice, is distinctive due to its extremely low SiO_2 mineral content, primarilly constituted by quartz and, subordinately, tridymite. Similar to cristobalite, tridymite reflects the mineralization processes in high-temperature (between 870 °C and 1470 °C) and low-pressure environments. Smectite content in this rock is practically absent, while the high amorphous content (18.1 wt%) suggests a substantial cryptocrystalline component within the groundmass of the sample. The quartz–trachyte OD_05 shows a markedly higher quartz content than the previous samples, with high-temperature SiO_2 polymorphs notably absent. A distinctive feature of the M. San Daniele trachyte is its feldspar composition, where sanidine predominates over plagioclase and anorthoclase, as confirmed by petrographic studies. Mineralogically, the M. San Daniele quartz–trachyte is most similar to the two rhyolite samples OD_13 and OD_14. However, these rhyolites are characterized by higher quartz concentrations than the San Daniele sample OD_05, exceeding 20 wt%. Additionally, these two samples show a slightly lower sanidine concentration than OD_05, with a noticeable increase in Na- and Ca-rich feldspars of the anorthoclase (13.2–18.7) and plagioclase types, respectively. This mineralogy closely matches that of a geological sample from the Forche del Diavolo outcrop reported in Dilaria et al. [40] (Figure 10).

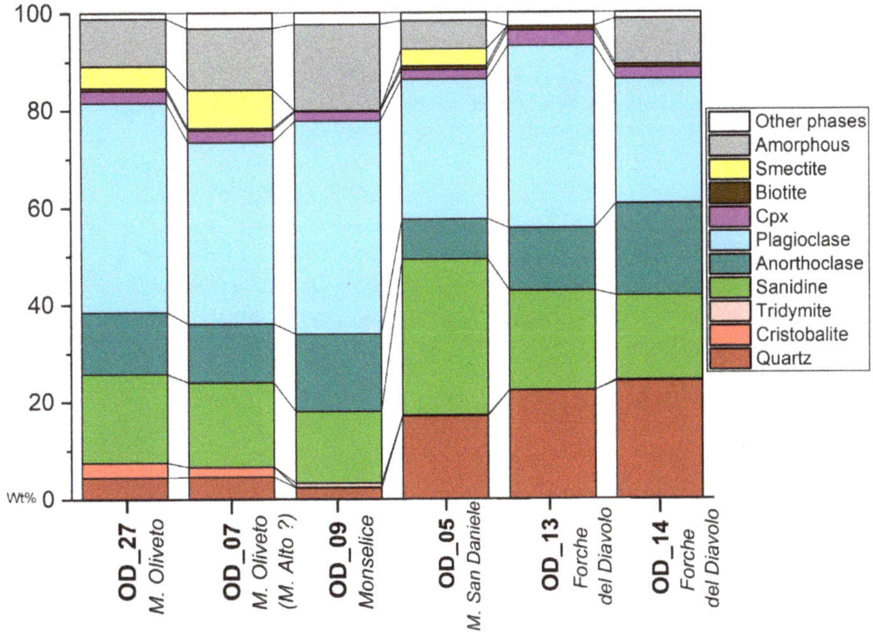

Figure 10. Cumulative histogram bars of descriptive mineral phases of the archeological samples from Oderzo among the Euganean volcanic rocks analyzed via QPA-XRPD.

5. Discussion and Conclusions

5.1. Stone Provenance

The data obtained from the analyses carried out on the collected samples provide a fairly complete picture of stone use in Roman Oderzo. Despite the different levels of detail that we have been able to obtain for determining the provenance of volcanic and sedimentary rocks, we can state with a good degree of certainty that stones were imported from almost all the main stone basins exploited in ancient times for the construction of structures and infrastructure in the city, as well as for the production of architectural artifacts (Table 2).

Most of the samples are sedimentary rocks, primarily from the eastern part of the region, specifically Aurisina limestone (nine samples), possibly Trieste Karst limestone (five samples), and Istrian stone (three samples). However, rocks from the western part of the region are also described, particularly from the Berici Hills, where Vicenza stone is quarried (three samples). The only samples that cannot be unequivocally attributed to a specific outcrop are those of Scaglia Rossa (six samples) and sandstone (four samples), whose formation outcrops in several parts of the region, including the prealpine area, which is only 30 km away from Oderzo. The relatively short distance from the outcrop area, which could have been easily reached by traveling up the Piave Valley, might suggest the import of Scaglia Rossa from this area located to the north of the town. However, at this stage of the study, this can only be a hypothesis as there is still a lack of petrographic elements that would allow for the archeological samples to be assigned to such a basin.

On the other hand, 16 of the analyzed samples are fully compatible with the magmatic outcrops of the Euganean Hills. The application of the multi-analytical approach described above allowed us to identify the specific quarries of origin with a good degree of accuracy, namely M. Oliveto (eight samples), Monselice (four samples), M. Alto (one sample), and M. San Daniele (one sample). However, for two samples, which differ from the others because of their rhyolitic chemistry, we can only predict their potential outcrops of provenance since the our current in-house database of geological rhyolite samples is not detailed enough to be considered comprehensive of all the different possible rhyolite sources in the Euganean territory. However, from a petrographical, mineralogical, and geochemical point of view, the samples seem comparable with the geological ones collected from this outcrop and reported in Dilaria et al. [40]. Therefore, according to the current state of research, these samples were associated with the Euganean outcrop of Forche del Diavolo.

In summary, the builders of Opitergium obtained stone material from both the western (Euganean Hills and Berici Hills) and eastern (Trieste Karst and Istria) regions. There remains a fraction of elements (Scaglia Rossa, sandstone) that cannot be easily attributed to a single quarrying basin, but its petrographic profile is nevertheless compatible with prealpine rock outcrops.

5.2. *Ways of Use and Chronology*

The identification of the stones used in structures and for the production of architectural elements suggests some interesting considerations about the association of particular lithotypes with specific ways of use (Figure 11).

One immediately obvious fact is that Euganean trachyte was used almost exclusively for the construction of infrastructure. For instance, the cover of the southern drainage channel of the forum and the two paving stones sampled in the area of the Ex Carceri are made of Euganean trachyte. The contexts in which trachyte appears to be most widely used are in the forum, whose paving, confirming what had been already stated by Zara [3], is primarily made of trachyte from M. Oliveto, and the city gate, where trachyte blocks from Monselice, M. Oliveto, and M. San Daniele were used for paving the *cavaedium*. The co-presence of stones from different quarries in the same pavement, which is actually an extension of the paved road that crossed the walls at the city gate, implies the simultaneous supply of material from different quarries, a reasonable solution to meet the large demand for stones needed to pave the whole road network of the city [3].

It seems clear that the ancient builders of *Opitergium* preferred, or were constrained by trade agreements whose nature is still unclear, to obtain trachyte exclusively from quarries located in the eastern and southeastern parts of the Euganean Hills. It is interesting to note the absence of materials from the M. Merlo quarry, which on the contrary is the most commonly used variety in Padua and Vicenza [3]. Outcrops from the western and southwestern slopes of the Euganean Hills are also absent.

The two samples of rhyolite taken from two pillars located on the side of the paved road in the area of the Ex Carceri deserve special attention. In this case, the builders' choice of material could have been influenced by the morphology of the outcrop itself. At Forche del Diavolo, in fact, located in the western part of the Euganean Hills, the rhyolite exhibits a natural columnar structure (Figure 12).

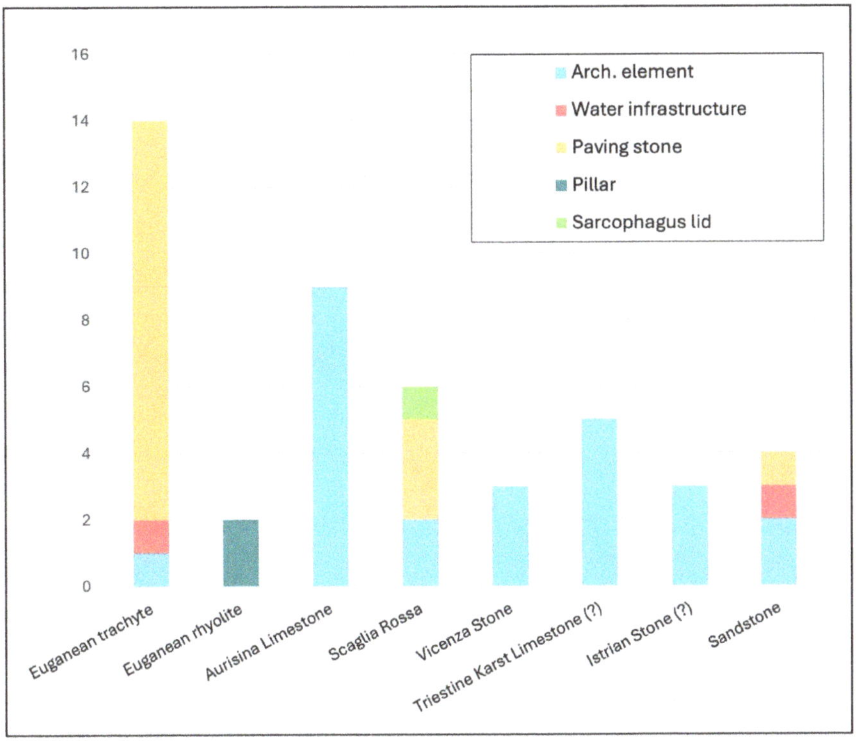

Figure 11. Diagram showing the correlation of documented lithotypes with specific functions.

Figure 12. Rhyolitic outcrop of Forche del Diavolo (photo by S. Dilaria).

Although used in smaller quantities than volcanic rocks, sedimentary rocks are also documented in infrastructures, primarily for restoration works. For example, between the 3rd and 4th centuries, the floor of the *cavaedium* of the city gate was restored using slabs and blocks of Scaglia Rossa, and a threshold of Aurisina limestone was also laid. Additionally, the restoration of the forum square, which involved replacing some slabs, was carried out using sedimentary rocks, namely sandstone.

However, the main use of sedimentary rocks appears to be in the production of architectural elements (Figure 13). There is a noticeable preference for Vicenza stone in the creation of columns and capitals, likely due to its workability and esthetic qualities. In contrast, Aurisina limestone is favored for architraves and cornices.

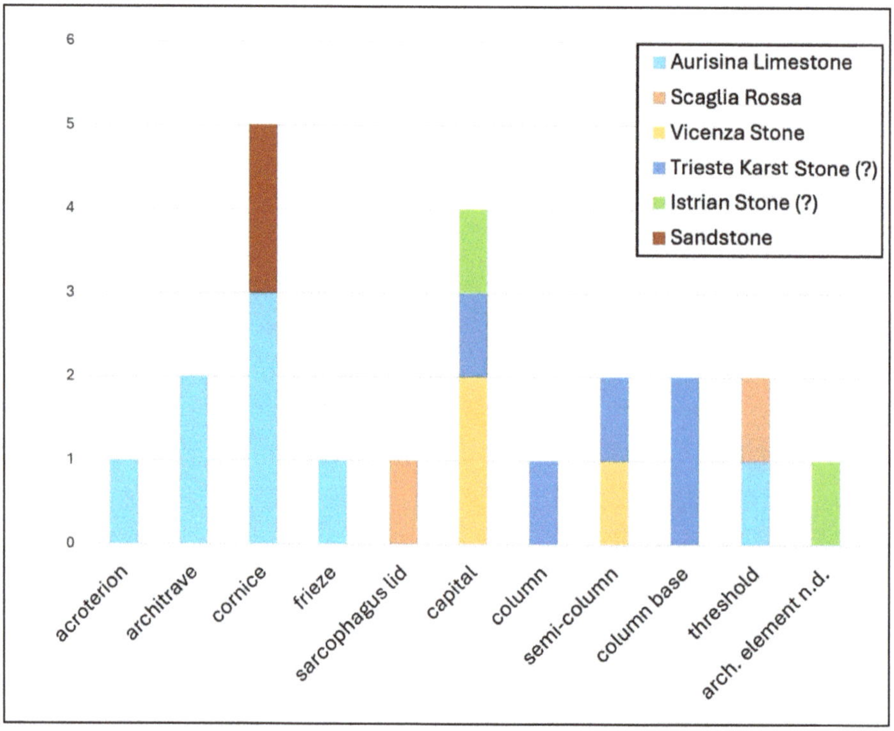

Figure 13. Diagram showing the correlation of documented sedimentary stones with specific architectural elements.

Regarding chronology, of the samples twenty, nineteen of which are architectural elements, while one was a sarcophagus lid that was reused in the 7th century wall; we do not have precise chronological data. The remaining 16 samples, however, belong to in situ structures in the area of the forum and the Ex Carceri, and this allows us to make some observations about the chronology of the use of certain lithotypes.

By analyzing the available data, we can conclude that between the second half of the 1st century BC and the 1st century AD, a large quantity of stone material was imported to the sites of *Opitergium*, particularly Euganean trachyte and, to a lesser extent, Aurisina limestone and sandstone. This finding is not surprising given that the paving of the *forum*, as well as the paving of the road and the *cavaedium* in the Ex Carceri, dates back to the transition between the 1st century BC and the 1st century AD.

According to the study of the samples, after this period Oderzo would not have imported new trachyte, or at least not in large quantities, until at least the 3rd century AD, an important phase of infrastructural restructuring in the Ex Carceri area is dated. It is also in this period that we find the most significant examples of Scaglia Rossa, along with the only two instances of rhyolite used in the construction of two pillars placed along the paved road.

However, no particular trends seem to emerge in the importation of Aurisina limestone, which is found in structures and used for the production of artifacts starting from the 1st century AD and continuing at least until the 4th century AD.

5.3. Trade Networks

The identification of the quarries of origin for the stones used in Oderzo allows us to expand our considerations to the broader topic of the dynamics of stone material circulation in Roman times.

Like today, the region was crossed by a dense network of waterways, which served as a communication system between urban centers, the coast, and, of course, the aforementioned quarries. It was mainly along these waterways that heavy loads of stone extracted from the quarries were transported to urban construction sites, taking advantage of the numerous roads that crossed the region only when strictly necessary. To reconstruct which of these routes were used to bring stone resources to Oderzo, it is necessary to recall the main features of the *X Regio*'s hydrographic and road network (Figure 14).

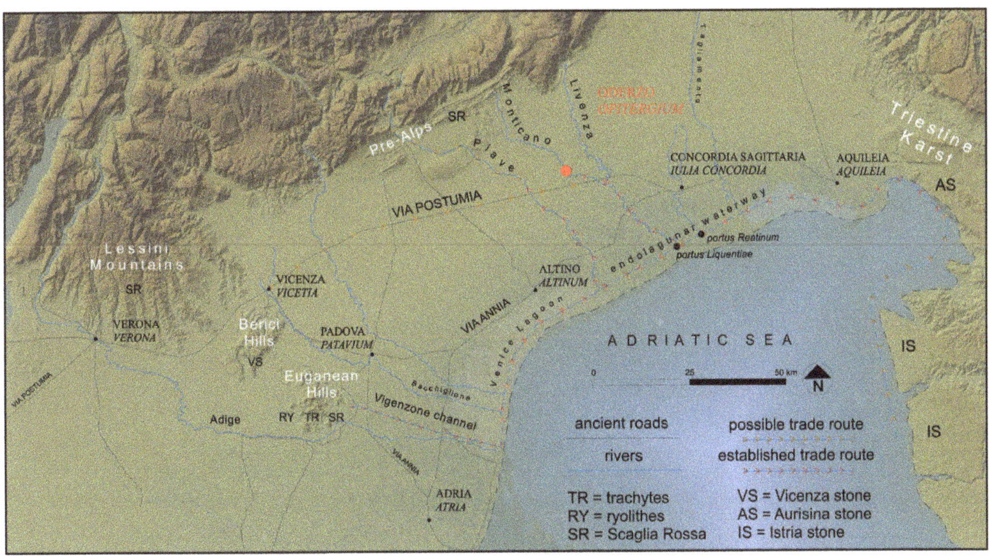

Figure 14. Map of the *X Regio* depicting the extraction basins and trade networks used by ancient Oderzo for stone supply (modified from [40]; basemap elaboration by P. Kirschner, J. Bonetto).

Major rivers, the most important of which are the Po, the Adige, the Brenta, the Piave, the Livenza, and the Tagliamento, flow through what is now the Veneto region. Numerous tributaries and smaller rivers traversed the lower plain, surrounding the main urban settlements. These rivers, along with the larger ones, flowed into a complex endolagunar waterway route, which was formed by artificial canals and lagoons, which connected Ravenna to Aquileia [57]. The development of this system dates back to the transition between the 1st century BC and the 1st century AD, running roughly parallel to the *Via Annia*, which had been constructed more than a century earlier [26,58]. A little further north was the already mentioned *Via Postumia*, which crossed the entire Transpadana region, connecting Genoa with Aquileia and passing just south of Oderzo.

The close connection between river and land routes created a capillary communication and trade system in which urban centers such as Oderzo served as commercial crossroads. It is not surprising then that building materials, especially stone ones, could be brought to the city's construction sites even from geographically distant areas.

The most frequently identified building material among the analyzed samples is, as mentioned above, limestone which came from both the Berici Hills and the Trieste Karst and possibly the Istrian peninsula. Given the short distance of the quarries from the coast [1] and the economic advantages of transport by sea [59], it is likely that the material from the Karst and Histria regions reached Oderzo by sea, possibly using the aforementioned articulated endolagunar system to reach the *portus Liquentiae* [60]. From there, it was possible to go up the Livenza river and then the Monticano river until finally reaching *Opitergium*.

Different possible solutions can be proposed for the transportation of the soft limestone from the Berici Hills. It is possible, if not very likely, that the Bacchiglione river was used to transport the quarried stone, a widespread practice as shown by the discovery of a wreck and its load of architectural elements [20]. In fact, the river flowed by the northeastern sector of the hills and then toward the Venetian lagoon, where it could join the endolagunar waterway. From there, the load could continue eastward until it reached the mouth of the Piave river, which is near today's small coastal town of Jesolo [60]. The stone load could then travel up the river until it reached the area nearby Oderzo, where it would finally be transported by land for the last leg of the journey. However, the Piave route is not the only possible option; it is also likely that the stone continued its journey to the *portus Liquentiae* following the route hypothesized for the limestones coming from the Karst region. Such a route would have allowed for closer access to the urban center by taking advantage of waterways and minimizing overland travel. However, a third hypothesis should be considered, although it is less likely given the economic disadvantages associated with it [59]. To the northwest of the Berici Hills ran the aforementioned *Via Postumia*, which passed through Vicenza and continued eastward until it reached Oderzo's territory, passing just south of the city. Therefore, the limestone quarried in the Berici Hills could have been transported to Vicenza and from there to Oderzo by charts. This remains a viable hypothesis, especially considering the relatively low demand for such materials; this seems likely given the analysis of the archeological samples examined in this study.

On the other hand, it seems reasonable to exclude the hypothesis of a land transport route for the loads of trachyte coming from the quarries of M. Oliveto, M. Alto, and Monselice given the large quantity of material that was surely needed to pave the forum and the city roads. Indeed, the route suggested by J. Bonetto [40], who identifies the Vigenzone canal as the ancient *Togisonus* mentioned by Pliny the Elder (*Naturalis Historia*, 3.121), appears convincing. The canal serves as a link between the eastern slope of the Euganean district, where M. Oliveto and M. Alto are located, and the endolagunar waterway already mentioned several times. From there, the stone load could have been transported eastward along the Adriatic arc to the *Liquentiae* port and finally to Oderzo by following the same routes proposed for the limestones. The Monselice trachyte, on the other hand, could have reached the endolagunar way via a palaeo-branch of the Adige river, which flowed near the southern sector of the Euganean Hills.

The Scaglia Rossa likely followed a similar route, granted it was quarried from the Euganean Hills or the Lessini Mountains near which the Adige river flowed. If it came from the Prealps, however, it probably followed the course of the Piave river, which flows through present-day Valdobbiadene and Possagno, passing by the Montello and descending to the plain close to the town. The sandstone, too, could have been transported along this route if it was quarried from the Miocene outcrops that define the foothills between Bassano del Grappa and Vittorio Veneto [51].

This network of waterways and routes clearly played a strategic role in the trade system of the region and served as a vital source of raw materials, including stone, for urban development. The variety of lithotypes found in construction, even from quarries located some distance away, further highlights the significance of the ancient settlement and defines *Opitergium* as a grand, monumental city.

Having analyzed the uses and trade routes of the different lithotypes documented in Oderzo, we can now broaden our discussion to the reasons that compelled the builders to choose a specific stone outcrop. The city, as already mentioned, is situated in the heart of the Treviso–Venetian plain, at a point that is equidistant from both the western and eastern outcrops, about 80–90 km from the Euganean Hills and the neighboring Berici Hills, while the distance to the Trieste Karst is about 90–100 km. Therefore, there are no quarrying basins, except perhaps those along the Prealps, which are only 30 km away from Oderzo, that are particularly close to the city for which one could assume a clear preference.

It was other reasons then, rather than proximity to the quarries, that dictated the selection of the different lithotypes documented. It is reasonable to assume that the first factor guiding the builders' choices must have been the physical and mechanical properties in the rock that made it particularly suited to the function it was destined to. Trachyte, for example, is a rock that is relatively easy to quarry and work with, and it is nevertheless characterized by good hardness, resistance to impact and compression, and poor polishability by friction [3], characteristics that make it a particularly suitable material for use as a floor covering, as found, for example, in the Forum of Oderzo.

The Aurisina limestone is also one of the most suitable carbonate formations for building. In fact, it is a compact limestone that is resistant to use and compression [49], qualities that were certainly already well known in Roman times. Therefore, it is not surprising that in Oderzo, most of the architectural elements were made of Aurisina limestone. It is also an esthetically pleasing material that, when polished, can resemble marble, which may have led the ancient builders to choose it as a viable and cheaper alternative. On the other hand, Vicenza stone seems to be less durable. However, its lower hardness and compactness make it easy to work, which was undoubtedly an advantage for architectural decoration.

The main physical–mechanical properties of each lithotype were certainly well known to both the builders and the quarry workers, between whom there must have been a close professional and collaborative relationship [59]. It is therefore likely that the choice of stone material to be used was not suggested by the client but by the very workmen charged with completing the building project. Similarly, it cannot be ruled out that there were close, albeit undefined, links between the client and/or architect of a particular building and the source of the material used on that site.

6. Conclusions

The variety of lithotypes found in construction, even from quarries located some distance away, further highlights the significance of the ancient settlement and defines *Opitergium* as a grand, monumental city.

The builders of *Opitergium* sourced stone materials from both the western regions (Euganean Hills and Berici Hills) and the eastern regions (Trieste Karst and Istria). A portion of the materials, including Scaglia Rossa and sandstone, cannot be definitively linked to a specific quarrying area, though their petrographic characteristics are consistent with prealpine rock formations.

The identification of the stones used in structures and for the production of architectural elements suggests a clear prevalence of Euganean trachyte for infrastructure and sedimentary rocks for the production of architectural elements. We can also affirm that the considerable use of stone in the building sites of Oderzo dates back to the transition between the 1st century BC and the 1st century AD, a time frame that aligns perfectly with the onset of the major urban and monumental projects that defined Oderzo and the entire *X Regio*.

The transport of material from the quarries to the urban centers probably took place mostly along the river routes that, both then and now, characterized the Venetian plain. This network of waterways and routes clearly played a strategic role in the trade system of the region and served as a vital source of raw materials, including stone, for urban development.

Supplementary Materials: The following supporting information can be downloaded at the following website: https://www.mdpi.com/article/10.3390/heritage8020044/s1, Supplementary Table S1: Results of the XRF analyses reported as wt% in oxides for major elements and as ppm for trace elements. B.d. = below detection limit. Supplementary Table S2: Classification function coefficients for trachytes and ryolites of the Euganean volcanic districts (markers) considered in this study. Supplementary Table S3: Results of the QPA-XRPD analysis of selected quarry-correlated archeological samples of volcanic rocks. b.d. = below detection limit.

Author Contributions: Conceptualization: C.G., C.P. and S.D.; methodology: C.P., S.D. and C.M.; formal analysis: C.G., S.D. and C.M.; investigation: C.G., S.D., and C.M.; writing: C.G. (Section 1; Section 2; Section 3.1; Section 3.2; Section 5), C.P. (Section 2; Section 3.1; Section 5), S.D. (Section 3.2; Section 4.2), C.M. (Section 4.1) and J.B. (Section 5). All authors have read and agreed to the published version of the manuscript.

Funding: This research was supported in the framework of the PON REACT-EU 2014/2020 projects DM 1061/2021—Action IV.6, BUILDing sustainABILITY in the Ancient World. Urban development, land use and conservation of the ecosystem and DM 1061/2021—Action IV.5—Cities, architecture and environmental sustainability: selection, exploitation and management of natural resources for the construction of urban centers in Roman Cisalpine. The preparation of the samples for OM, XRF and XRPD analyses was developed at the Department of Geosciences of the University of Padova. The research was implemented and funded within the scopes of the University of Padova under the World Class Research Infrastructures (WCRI) program—SYCURI (Synergic Strategies for Culture Heritage at Risk). The paper was also financed from the project "Trade and use of volcanic pozzolans in the Roman world. A natural material for the production of eco-sustainable concrete of antiquity" (Principal investigator: Simone Dilaria, BIRD 2023 of the Department of Cultural Heritage of the University of Padova, project code BIRD230232/23. Stone samples were collected with the authorization of the Soprintendenza Archeologia, Belle Arti e Paesaggio per l'area metropolitana di Venezia e le province di Belluno, Padova e Treviso (Prot. n. 23289 del 15 July 2022).

Data Availability Statement: The data other than those reported in the text and in the supplementary materials are available on request from the corresponding authors.

Conflicts of Interest: The authors declare no conflicts of interest.

References

1. Previato, C. Aquileia. In *Materiali, Forme e Sistemi Costruttivi Dall'età Repubblicana Alla Tarda età Imperiale*; Padova University Press: Padova, Italy, 2015.
2. Previato, C. Aurisina limestone in the Roman Age: From Karst Quarries to the cities of the Adriatic basin. In *AMOSIA XI. Interdisciplinary Studies of Ancient Stone. Proceedings of the Eleventh International Conference of ASMOSIA, Split 18–22 May 2015*; Poljak, M., Marasović, K., Eds.; University of Split: Split, Croatia, 2018; pp. 933–939.

3. Zara, A. *La Trachite Euganea: Archeologia e Storia di una Risorsa Lapidea del Veneto Antico*; Edizioni Quasar: Rome, Italy, 2018.
4. Previato, C.; Zara, A. A database and GIS project about the quarrying, circulation and use of stone during the roman age in Regio X—Venetia et Histria. The case study of euganean trachyte. In *AMOSIA XI. Interdisciplinary Studies of Ancient Stone. Proceedings of the Eleventh International Conference of ASMOSIA, Split 18–22 May 2015*; Poljak, M., Marasović, K., Eds.; University of Split: Split, Croatia, 2018; pp. 597–609.
5. Germinario, L.; Zara, A.; Maritan, L.; Bonetto, J.; Hanchar, J.M.; Sassi, R.; Siegesmund, S.; Mazzoli, C. Tracking trachyte on the Roman routes: Provenance study of Roman infrastructure and insights into ancient trades in northern Italy. *Geoarchaeology* **2018**, *33*, 417–429. [CrossRef]
6. Dilaria, S.; Ricci, G.; Secco, M.; Beltrame, C.; Costa, E.; Bonetto, J.; Artioli, G. Vitruvian binders in Venice: First evidence of Phlegraean pozzolans in an underwater Roman construction in the Venice Lagoon. *PLoS ONE* **2024**, *19*, e0313917. [CrossRef] [PubMed]
7. Cornale, P.; Rosanò, P. *Le Pietre Tenere del Vicentino. Uso e Restauro*; La Grafica & Stampa: Vicenza, Italy, 1994.
8. Lazzarini, L. Pietra d'Istria: Genesi, proprietà e cavatura della pietra di Venezia. In *La Pietra d'Istria e Venezia. Atti del Seminario di Studio*; Fiorentin, N., Ed.; Cierre Edizioni: Verona, Italy, 2006; pp. 24–45.
9. Astolfi, G.; Colombara, F. *La Geologia dei Colli Euganei*; Programma: Padova, Italy, 1990.
10. Guerra, R. *Liquentia: Un Fiume Nella X Regio Augustea Venetia et Histria*; Tipolitografia Rubino: S. Stino di Livenza, Italy, 2005.
11. Buonopane, A. Estrazione, lavorazione e commercio dei materiali lapidei. In *Il Veneto nell'Età Romana*; Buchi, E., Ed.; Banca Popolare di Verona: Verona, Italy, 1987; Volume 1, pp. 187–224.
12. Bridi, E. L'edilizia Pubblica di Verona Romana: Materiali, Cantieri, Sistemi Costruttivi, Economie e Committenze. Ph.D. Thesis, Università Degli Studi di Padova, Padova, Italy, 2023.
13. Capedri, S.; Grandi, R.; Venturelli, G. Trachytes used for paving Roman roads in the Po Plain: Characterization by petrographic and chemical parameters and provenance of flagstones. *J. Archaeol. Sci.* **2003**, *30*, 491–509. [CrossRef]
14. Flügel, E.; Flügel, C. Applied microfacies analysis: Provenance studies of Roman mosaic stones. *Facies* **1997**, *37*, 1–48. [CrossRef]
15. Roffia, E.; Bugini, R.; Folli, L. Stone materials of the Roman villas around Lake Garda (Italy). In *ASMOSIA VII*; Maniatis, Y., Ed.; BCH Supplement 51; Ecole Francaise d'Athenes: Athens, Greece, 2009; pp. 559–570.
16. De Vecchi, G. Analisi delle pietre e dei materiali edilizi. In *L'Area del Capitolium di Verona. Ricerche Storiche e Archeologiche*; Cavalieri Manasse, G., Ed.; Cortella Poligrafica: Verona, Italy, 2008.
17. Falezza, G.; Zorzin, R.; Lazzarini, L.; Bonetto, J. Il contesto geologico e il materiale da costruzione impiegato nel santuario. In *Archeologia e Storia sul Monte Castelon di Marano di Valpolicella*; Bruno, B., Falezza, G., Eds.; Sap Società Archeologica S.R.L.: San Lorenzo, Italy, 2016; pp. 233–244.
18. Lazzarini, L. The minero-petrographic analysis of two flagstones ("basoli") recovered from Treporti (Venice Lagoon) and Altinum. In *Crossing the Water. The Venice Lagoon from Antiquity Throughout the Centuries*; Bassani, M., D'Acunto, G., Madricardo, F., Eds.; L'Erma Di Bretschneider: Roma, Italy, 2022; pp. 155–161.
19. Brajkovič, R.; Djurić, B.; Gerčar, D.; Cvetko Tešović, B.; Rožič, B.; Gale, L. Mid-Cretaceous calcarenite in stone products from the Roman colony of Emona, Regio X (modern Ljubljana, Slovenia). *Archaeometry* **2023**, *65*, 17–35. [CrossRef]
20. Previato, C.; Zara, A. Il trasporto della pietra di Vicenza in età romana. Il relitto del fiume Bacchiglione. *Marmora* **2014**, *10*, 59–78.
21. Bonetto, J.; Dilaria, S.; Girotto, C.; Previato, C.; Falezza, G.; Facchi, A. Tra le cave e il Delta del Po: Vie di trasporto e dinamiche del commercio della pietra. In *Forme e Aspetti del Paesaggio. Percorrendo le Strade dell'Italia Antica*; Atlante Tematico di Topografia Antica; L'Erma di Bretschneider: Roma, Italy, 2025.
22. Tirelli, M. Oderzo. In *Il Veneto nell'Età Romana. II. Note di Urbanistica e di Archeologia del Territorio*; Cavalieri Manasse, G., Ed.; Banca Popolare di Verona: Verona, Italy, 1987; pp. 358–390.
23. Busana, M.S. *Oderzo. Forma Urbis. Saggio di Topografia Antica*; "L'Erma" di Bretschneider: Roma, Italy, 1995.
24. Tirelli, M. *Itinerari Archeologici di Oderzo*; Canova: Treviso, Italy, 2003.
25. Tirelli, M. Opitergium, municipio romano. In *L'anima delle Cose. Riti e Corredi della Necropoli Romana di Opitergium. Catalogo della Mostra*; Mascardi, M., Tirelli, M., Eds.; Ca' Foscari-Digital Publishing: Venezia, Italy, 2019; pp. 27–36.
26. Bosio, L. *Le Strade Romane della Venetia e dell'Histria*; Esedra: Padova, Italy, 1997.
27. Tirelli, M. L'area del foro di Oderzo (Treviso). In *La Città nella Città. Sistemazione di RESTI Archeologici in Area Urbana: L'Italia del Nord*; Croce Da Villa, P., Dal Pos, M., Penzo, A., Eds.; Tipografia Sagittaria: Concordia Sagittaria, Italy, 1989; pp. 39–47.
28. Tirelli, M.; Sandrini, G.M. Oderzo. Balneum nei quartieri settentrionali. *Quad. Archeol. Veneto* **1990**, *6*, 155–161.
29. Ruta Serafini, A.; Tirelli, M. Dalle origini all'alto medioevo: Uno spaccato urbano di Oderzo dallo scavo dell'ex stadio. *Quad. Archeol. Veneto* **2004**, *20*, 135–152.

30. Tirelli, M.; Malizia, A. Note preliminari sul rinvenimento di domus romane nel settore urbano nord-orientale dell'antica Oderzo. *Quad. Archeol. Veneto* **1985**, *1*, 151–165.
31. Tirelli, M. La domus di via Mazzini ad Oderzo (Treviso). *Quad. Archeol. Veneto* **1987**, *3*, 171–192.
32. Tirelli, M. Oderzo: Resti di basolato stradale tra Piazza Vittorio Emanuele e Piazza Castello. *Quad. Archeol. Veneto* **1989**, *5*, 75–76.
33. Cipriano, S.; Sandrini, G.M. La banchina fluviale di Opitergium. *Antich. Altoadriatiche* **2001**, *46*, 289–294.
34. Bonetto, J. *Mura e Città nella Transpadana Romana*; Fondazione "Antonio Colluto": Portogruaro, Italy, 1998.
35. Braconi, M. Il banchetto e la caccia su due mosaici pavimentali di Oderzo fra tradizione iconografica e autorappresentazione. *Antich. Altoadriatiche* **2016**, *86*, 281–303.
36. Castagna, D.; Tirelli, M. Evidenze archeologiche di Oderzo tardo-antica ed altomedievale: I risultati preliminari di indagini recenti. In *Città, Castelli, Campagne nei Territori di Frontiera (Secoli VI–VII)*; Brogiolo, G.P., Ed.; Documenti di Archeologia; Sap Società Archeologica S.R.L.: San Lorenzo, Italy, 1995; pp. 121–134.
37. Tirelli, M. Spolia dalla necropoli opitergina: Monumenta. In *La Necropoli di Opitergium, Atti della Giornata di Studi Intorno alla Mostra L'anima delle Cose, Oderzo, 25 Maggio 2021*; Mascardi, M., Tirelli, M., Vallicelli, M.C., Eds.; Edizioni Ca' Foscari: Venezia, Italy, 2023; pp. 11–30.
38. Girotto, C.; Marchet, B. Pietre e marmi antichi: Forme del riuso nell'area delle Ex Carceri di Oderzo (TV). In *Atti del Convegno Forme in Movimento. Modelli, Metodi e Contesti tra Continuità e Innovazione*; Padova University Press: Padua, Italy, 2025; *in press*.
39. Forlati Tamaro, B. *Iscrizioni Lapidarie Latine del Museo Civico di Oderzo*; Marton Editore: Treviso, Italy, 1976.
40. Dilaria, S.; Bonetto, J.; Germinario, L.; Previato, C.; Girotto, C.; Mazzoli, C. The stone artifacts of the National Archaeological Museum of Adria (Rovigo, Italy): A noteworthy example of heterogeneity. *Archaeol. Anthropol. Sci.* **2024**, *16*, 14. [CrossRef]
41. Calvino, F. *Le Cave dei Colli Euganei. Caratteristiche Geominerarie. Importanza Economica. Prospettive di Sviluppo*; Consorzio per la Valorizzazione dei Colli Euganei: Padova, Italy, 1967.
42. Germinario, L.; Hanchar, J.M.; Sassi, R.; Maritan, L.; Cossio, R.; Borghi, A.; Mazzoli, C. New petrographic and geochemical tracers for recognizing the provenance quarry of trachyte of the Euganean Hills, northeastern Italy. *Geoarchaeology* **2018**, *33*, 430–452. [CrossRef]
43. Rietveld, H.M. Line profiles of neutron powder-diffraction peaks for structure refinement. *Acta Crystallogr.* **1967**, *22*, 151–152. [CrossRef]
44. Dollase, W. Correction of intensities for preferred orientation in powder diffractometry: Application of the March Model. *J. Appl. Crystallogr.* **1986**, *19*, 267–272. [CrossRef]
45. Taut, T.; Kleeberg, R.; Bergmann, J. The new Seifert Rietveld program BGMN and its application to quantitative phase analysis. *Mater. Struct.* **1998**, *5*, 55–64.
46. Cucchi, F.; Piano, C. *Brevi Note Illustrative della Carta Geologica del Carso Classico Italiano*; Università degli Studi di Trieste: Trieste, Italy, 2013.
47. Cucchi, F.; Biolchi, S.; Zini, L.; Jurkovsek, B.; Kolar Jurkovsek, T. Geologia e geomorfologia del Carso Classico. In *Le Acque del Carso Classico. Progetto Hydrokarst/Vodonosnik Klasičnega Krasa*; Projekt Hydrokarst; EUT Edizioni: Trieste, Italy, 2015; pp. 23–52.
48. Carulli, G.B.; Onofri, R. *I Marmi del Carso*; Del Bianco: Udine, Italy, 1969.
49. Cucchi, F.; Gerdol, S. *I Marmi del Carso Triestino*; Camera di Commercio e Industria: Trieste, Italy, 1985.
50. Massari, F.; Medizza, F. Stratigrafia e paleogeografia del Campaniano e Maastrichtiano nelle Alpi Meridionali (con particolare riguardo agli hard grounds della Scaglia rossa veneta). *Mem. Ist. Geol. Mineral. Dell'Università Padova* **1963**, *28*, 1–64.
51. Antonelli, R.; Barbieri, G.; Dal Piaz, G.V.; Dal Pra, A.; De Zanche, V.; Grandesso, P.; Mietto, P.; Sedea, R.; Zanferrari, A. *Carta Geologica del Veneto 1:250,000: Una Storia di Cinquecento Milioni di Anni*; Regione del Veneto: Venezia, Italy, 1990.
52. Cattaneo, A.; De Vecchi, G.P.; Menegazzo Vitturi, L. *Le Pietre Tenere dei Colli Berici*; Società Cooperativa Tipografica: Padova, Italy, 1976.
53. Le Bas, M.J.R.; Maitre, R.L.; Streckeisen, A.; Zanettin, B. IUGS Subcommission on the systematics of igneous rocks, a chemical classification of volcanic rocks based on the total alkali-silica diagram. *J. Petrol.* **1986**, *27*, 745–750. [CrossRef]
54. Germinario, L.; Cossio, R.; Maritan, L.; Borghi, A.; Mazzoli, C. Textural and mineralogical analysis of volcanic rocks by µ-XRF mapping. *Microsc. Microanal.* **2016**, *22*, 690–697. [CrossRef] [PubMed]
55. Maritan, L.; Mazzoli, C.; Sassi, R.; Speranza, F.; Zanco, A.; Zanovello, P. Trachyte from the Roman aqueducts of Padua and Este (north-east Italy): A provenance study based on petrography, chemistry and magnetic susceptibility. *Eur. J. Min.* **2013**, *25*, 415–427. [CrossRef]
56. Whitney, D.L.; Evans, B.W. Abbreviations for names of rock-forming minerals. *Am. Mineral.* **2010**, *95*, 185–187. [CrossRef]
57. Rosada, G. La direttrice endolagunare e per acque interne nella decima regio maritima: Tra risorsa naturale e organizzazione antropica. In *La Venetia nell'Area Padano-Danubiana. Le vie di Comunicazione, Atti del Convegno Internazionale, Venezia, 6–10 Aprile 1988*; CEDAM: Padova, Italy, 1990; pp. 153–182.
58. Uggeri, G. La nuova Via Annia da Roma ad Aquileia (153 a.C.). *J. Anc. Topogr.* **2012**, *22*, 157–158.

59. Russell, B. *The Economics of the Roman Stone Trade*; Oxford Studies on the Roman Economy; Oxford Academic: Oxford, UK, 2013.
60. Croce Da Villa, P. Concordia e Caorle. *Antich. Altoadriatiche* **1990**, *36*, 329–332.

heritage

MDPI

Article

X-CT Reconstruction as a Tool for Monitoring the Conservation State and Decay Processes of Works of Art and in Support of Restoration and Conservation Strategies

Laura Guidorzi [1,*], Alessandro Re [1,*], Francesca Tansella [1], Luisa Vigorelli [1], Chiara Ricci [2], Joseph Ryan [3] and Alessandro Lo Giudice [1]

1 Dipartimento di Fisica, Università di Torino & INFN—Sezione di Torino, Via Pietro Giuria 1, 10125 Torino, Italy; francesca.tansella@unito.it (F.T.); luisa.vigorelli@unito.it (L.V.); alessandro.logiudice@unito.it (A.L.G.)

2 Centro per la Conservazione ed il Restauro dei Beni Culturali "La Venaria Reale", Via XX Settembre 18, Venaria Reale, 10078 Torino, Italy; chiara.ricci@ccrvenaria.it

3 Research Institute for the Dynamics of Civilizations, Okayama University, 3-1-1 Tsushima-naka, Kita-ku, Okayama 700-8530, Japan; josephryan@okayama-u.ac.jp

* Correspondence: laura.guidorzi@unito.it (L.G.); alessandro.re@unito.it (A.R.)

Abstract: X-ray Computed Tomography (X-CT) is now an established technique for the investigation and diagnostics of Cultural Heritage. Its advantages include non-invasiveness, non-destructiveness, and the possibility of exploring the inner parts of an object without any modification. X-CT is often employed to investigate the construction methods of complex artifacts made with different parts or materials, but it is also able to support the analysis, intervention, monitoring and enhancement processes of artworks, creating digital models that can aid in the conservation and restoration procedures. In this work, several case studies are presented in which the CT technique has been decisive in identifying the effects of time and the events that occurred during the object's life influencing its state of conservation. These range from large objects, such as an 18th century CE writing cabinet or an ancient Egyptian wooden coffin, to very small artifacts, like Mesopotamian lapis lazuli beads or fragments of Roman colored glass. Additionally, the results obtained by μ-CT investigations on the conservation state of a bronze arrowhead uncovered from the Urama-chausuyama mounded tomb (Japan, Kofun period, end of the 3rd century CE) are presented here for the first time. Lastly, the versatility of the technique when applied with different setups is highlighted.

Keywords: X-ray computed tomography; preventive conservation; damage; xylophagous insects; corrosion; bronze arrowheads; Urama-chausuyama mounded tomb

1. Introduction

Many transformations can occur in a work of art, starting from its creation. Physical, chemical and biological changes can affect its structure, appearance, composition and integrity as time goes on. Sometimes changes are not visible from the outside, or conversely, an apparently destroyed object might actually hide inside a pristine part. In such cases, an analysis of the inner part of the object is needed to assess its actual conservation state. X-ray Computed Tomography (X-CT) has been borrowed from the medical field since the 1980s for the investigation and diagnostics of Cultural Heritage [1]. Its advantages include non-invasiveness, non-destructiveness and the possibility of exploring the inside of an object without modifying it in any way and without inducing activation. X-CT is based, like radiography, on the different X-ray attenuation coefficients characteristic of each material in relation to their composition, density or thickness, introducing in addition a complete rotation during the acquisition of the images: in this way, it allows the creation of 3D digital models to support the analysis, intervention, monitoring and enhancement processes of artworks. Moreover, the analysis can be repeated endless times at different stages in the life cycle of the artifact in order to monitor the evolution of the conservation state and identify an eventual decay of the constitutive material or of the structure. The possibility of performing the analysis in situ (e.g., museums) [2,3] facilitates the application of the technique even more, avoiding the displacement of the object outside its neutral environment and limiting any additional stress imposed on the materials. Conversely, large-scale X-CT can be a powerful technique for the non-destructive visual inspection of historic structures in our Cultural Heritage, such as damaged timber structural elements from an 18th century bridge [4].

In several cases, the technique is decisive in identifying the effects of time and the events that occurred during the object's life, influencing its state of conservation. Some examples are offered from the study of wooden-based objects, where X-CT is generally employed to detect cracks or deformations undermining the health of the artifact [4–6]. Dimensional changes can occur instead in the case of previously waterlogged artifacts when they are removed from the original environment: for shrinkage, collapse and cracks, even if the changes are very small, an evaluation can be successfully achieved by means of CT [7]. Mineralization of wood can also happen after prolonged contact between metallic and wooden parts of the same object, and a tomographic study can assess the stage of the process and help establish strategies for preventive conservation [8].

Also, the conservation conditions and environment can influence the preservation of the artifact, and X-CT scanning can be used to monitor the occurrence of features caused by the specific characteristics of the burial soil or the environmental conditions in museums [9–12]. In this regard, CT can also be exploited to analyze and extract the original form and scope of artifacts embedded in soil blocks that are heavily damaged from the time spent under burial, supporting restoration interventions [13–15].

The activity of biological factors can be surveyed as well by means of CT, which can, for example, reveal xylophagous insect strikes in wooden objects [16,17]; repeating the analysis some time after an anoxic treatment can be an easy way to confirm the success of the procedure, verifying that no new holes have been formed inside the artwork. The decay processes involving relatively modern materials can also be studied, such as cement mortars [18], which are constitutive components of architecture or objects that have become part of Cultural Heritage only in recent times.

Generally, X-ray CT can have limited investigative power for metallic artifacts due to the high absorption of this kind of radiation by heavy metals (iron, copper, tin, lead and so on). However, this limitation is not an issue for small and highly degraded metallic artifacts: if anything, the high contrast resulting from the difference in composition and density between the areas is helpful to better identify any remaining metal inside the object [19,20]. A quite common application of the technique is in numismatic studies, to be able to examine the metal hidden behind the corrosion layer and in some cases to extract the marks and decorations and successfully reach a classification of the coins [21–25]. In general, this also means that when cleaning the object is not possible, as it would also destroy the preserved part, X-CT guarantees at least the virtual valorization of the original artifact.

The examples of tomographic investigations reported in this work show how it is possible with versatile instrumentation to go beyond the effects of time and reveal the actual state of conservation of the object, while studying the causes that led to the eventual damage. After reporting specific time-dependent effects on artifacts already studied with

X-CT with different aims, the evaluation of metal corrosion in a bronze arrowhead from a Japanese mounded tomb is presented for the first time.

2. Materials and Methods

Most of the case studies presented here have been investigated in their entirety or with different purposes elsewhere, while in this work, we focus our attention on the effect of time-dependent damages or alterations on diverse materials. Specifically, the objects under investigation are (1) the "doppio corpo" writing cabinet by Pietro Piffetti [26], (2) the lid of Taiefmutmut's wooden coffin [27], (3) a soil block retrieved from the archaeological site of Villalfonsina (Abruzzo, Italy) [28], (4) a ceramic *tintinnabulum* [29], (5) lapis lazuli beads from the Royal Graves of Ur (Mesopotamia) [30], (6) fragments of colored glass from the archaeological site of Aquileia (North-Eastern Italy) [31], and (7) a historical wooden transverse flute (Traversiere) [32]. We refer to the cited literature for a complete description of the objects and their historical contexts. To take into account bulk metallic artifacts, we also report for the first time on a tomographic study on the state of conservation of a bronze arrowhead on display at the Okayama University Archaeological Museum in Japan. The arrowhead (number 12) is part of a collection of around 20 samples unearthed from the Urama-chausuyama mounded tomb (Kofun period, end of the 3rd century CE), located in Okayama Prefecture. Urama-chausuyama is a keyhole-shaped mounded tomb measuring 138 m in length. While its stone burial chamber was heavily looted and most of the burial goods had been lost, the 1988 excavation retrieved fragments of a bronze mirror, bronze and iron arrowheads, and single- and double-edged iron swords, in addition to tools such as axe heads and knives, agricultural implements such as sickles and spades, and fishing gear such as harpoons [33]. In particular, arrowhead number 12 displays a peculiar pentagonal shape (Figure 1). In recent years, the arrows were coated with a layer of paraloid B72 for conservation purposes. The surface, although cohesive in appearance due to a mineralization layer [34–36], shows relevant signs of degradation (localized to some areas and fractures on the surface), but inspection with optical microscopy could not provide information on the health conditions of the innermost part.

Several X-CT systems have been used for data acquisition, including both custom-made setups and benchtop machines. Custom setups are placed in shielded rooms as they can occupy a large amount of space, while benchtop ones deal with radioprotection employing a smaller closed cabinet. During the analysis, the rotation is always applied to the object, while the X-ray source and the detector are kept in the same position. The characteristics of different instrumentation and the chosen parameters for each analysis are reported in Table 1. Custom-made laboratory setups are based in two different locations; both systems have been developed by the Physics Department of the University of Torino and the INFN (Italian National Institute for Nuclear Physics). The micro-CT system employed is located at the Physics Department of the University of Torino, and it is thoroughly described in [32]. It is particularly of use when the investigation requires high resolution of very small details (the nominal minimum focal spot size is 5 μm, for a minimum voxel size of about 6–10 μm). The acquisition is rather quick thanks to the use of the available flat panel detectors, exploiting different active areas. The X-CT system for large objects is instead hosted at the Center for Conservation and Restoration "La Venaria Reale" in Venaria Reale, near Torino [37]. This setup can achieve a higher X-ray energy (source maximum voltage is 200 kV), but it relies on a linear detector horizontally scanning the area of interest to perform the two-dimensional (2D) radiographs [38]. The geometry of the setup and the X-ray beam characteristics were optimized for each case, taking into account the type of the material and the dimensions of the object. The different source-detector, source-object and object-detector distances can be modified, especially in the customized setup, to act on the magnification of the image. The projections were acquired through a custom-made Labview routine that synchronizes the rotation motion and the projection acquisition. The same routine was used to acquire "open-beam" images (same conditions, no sample) and "dark" images (X-rays off), which were averaged using MATLAB software (version R2023b) by MathWorks.

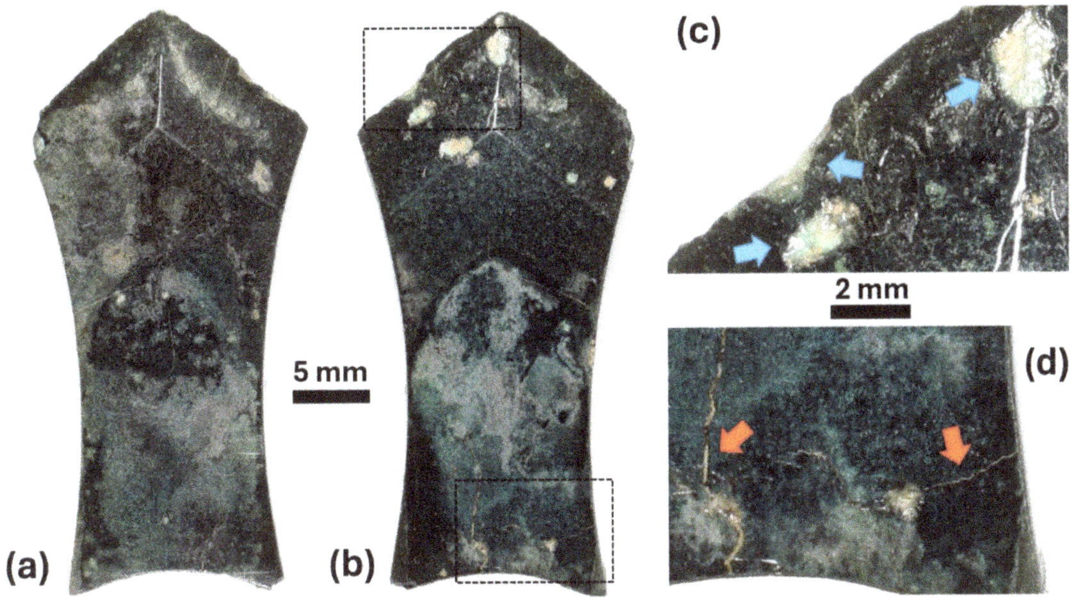

Figure 1. (**a**,**b**) Front and back side of the bronze arrowhead (sample 12) from the Urama-chausuyama mounded tomb, Okayama Prefecture, Japan. (**c**,**d**) Magnification of corroded parts, where material losses (blue arrows) and cracks (red arrows) are present.

The benchtop X-CT system used instead for the tomographic acquisition of the bronze arrowhead is a SHIMADZU inspeXio SMX-225CT FPD HR Micro Focus X-Ray CT System hosted at Shimadzu Techno-Research, Inc. in Kyoto, Japan. To reduce beam hardening effects during the analysis of metallic objects, a 200 kV tube voltage was set, and a 1 mm thick copper filter was used instead of an aluminum filter. The instrument is equipped with a microfocus X-ray source for high-resolution analysis and a flat panel to make the acquisitions faster (only 20 min for each acquisition). Taking also into account the geometric conditions given in Table 1, the voxel size obtained was 45 μm.

Table 1. Instrumentation and experimental parameters for the different X-CT acquisitions presented in this work.

Object	Piffetti's "Doppio Corpo"	Coffin Lid	Soil Block	*Tintinnabulum*	Lapis Lazuli Bead	Roman Glass	Flute	Bronze Arrowhead
Object dimensions	$129 \times 59 \times 312$ cm^3	$31 \times 50 \times 182$ cm^3	$40 \times 15 \times 10$ cm^3	$7.4 \times 11.2 \times 6.2$ cm^3	(ca.) $8.4 \times 3.5 \times 3.5$ mm^3	D_{max} 2.5 cm	$42.7 \times 3.4 \times 3.4$ cm^3	$3.5 \times 2.0 \times 0.5$ cm^3
X-ray source	General Electric Eresco 42MF4	General Electric Eresco 42MF4	General Electric Eresco 42MF4	General Electric Eresco 42MF4	Hamamatsu Microfocus L8121-03	Hamamatsu Microfocus L8121-03	Hamamatsu Microfocus L8121-03	SHIMADZU inspeXio SMX-225CT FPD HR Micro Focus X-Ray CT System
X-ray detector	Hamamatsu X-ray Line Sensor Camera C9750-20TCN	Hamamatsu X-ray Line Sensor Camera C9750-20TCN	Hamamatsu X-ray Line Sensor Camera C9750-20TCN	Hamamatsu X-ray TDI Camera C10650-321	Shad-O-Box 6K HS Flat Panel	Shad-O-Box 6K HS Flat Panel	Shad-O-Box 6K HS Flat Panel	
Rotary stage	Newport RV350PE	Newport RV350PE	Newport RV350PE	custom	Newport URS150BPP	Newport URS150BPP	Newport URS150BPP	
Tube voltage	180 kV	180 kV	200 kV	80 kV	150 kV	90 kV	90 kV	200 kV
Tube current	5 mA	5 mA	4.5 mA	10 mA	66 µA	111 µA	500 µA	120 µA
Focal spot size	3 mm	3 mm	3 mm	3 mm	7 µm	7 µm	50 µm	ca. 4 µm
Filter	Al (2 mm)	Al (2 mm)	Al (2 mm)	Al (2 mm)	Al (2 mm)	Al (2 mm)	Al (2 mm)	Cu (1 mm)
Detector pixel size	200 µm	200 µm	200 µm	48 µm	49.5 µm	49.5 µm	49.5 µm	130 µm
Detector pixels number	2560×1	2560×1	2560×1	4608×128	2304×2940	2304×2940	2304×2940	3000×3000
Detector active area	51.2×0.2 cm^2	51.2×0.2 cm^2	51.2×0.2 cm^2	22.1×0.6 cm^2	11.4×14.6 cm^2	11.4×14.6 cm^2	11.4×14.6 cm^2	40.6×40.6 cm^2
A/D converter	12 bit	12 bit	12 bit	12 bit	14 bit	14 bit	14 bit	16 bit
Detector scan speed (m/min) or integration time (s)	5 m/min	2.2 m/min	2 m/min	4 m/min	1.9 s	3 s	4 s	1 s
Rotation	360°	270°	270°	360°	360°	360°	360°	360°
Angular step	0.5°	0.25°	0.5°	0.5°	0.15°	0.15°	0.25°	0.3°
Number of projections	720 (×13 sections)	1080 (×6 sections)	540	720	2400	2400	1440 (×3 sections)	1200

Table 1. *Cont.*

Object	Piffetti's "Doppio Corpo"	Coffin Lid	Soil Block	*Tintinnabulum*	Lapis Lazuli Bead	Roman Glass	Flute	Bronze Arrowhead
Total time for CT scan	5.6 days	90 h	220 min	105 min	120 min	120 min	426 min	20 min
Source-Detector Distance (SDD)	2950 mm	3690 mm	2940 mm	1465 mm	650 mm	650 mm	1400 mm	800 mm
Source-Object Distance (SOD)	2140 mm	3180 mm	2640 mm	1368 mm	90 mm	100 mm	1300 mm	90 mm
Object-Detector Distance (ODD)	810 mm	510 mm	300 mm	97 mm	560 mm	550 mm	100 mm	710 mm
Magnification	1.3×	1.16×	1.11×	1.07×	7.22×	6.50×	1.08×	8.89×
Binning	4 × 4	2 × 2	none	2 × 2	none	none	none	3 × 3
Voxel size	580 µm	340 µm	180 µm	90 µm	7 µm	8 µm	46 µm	45 µm
Penumbra	1.1 mm	480 µm	340 µm	210 µm	44 µm	39 µm	4 µm	32 µm
Reconstruction	LLNL [1]	LLNL	LLNL	ParRec [2] + LLNL	ParRec	ParRec	ParRec	Instrumentation proprietary software
Visualization	VGStudio MAX 2.2	VGStudio MAX 2.2	VGStudio MAX 2.2	VGStudio MAX 2.2	ORS Dragonfly	ORS Dragonfly	ORS Dragonfly	MyVGL
Reference	[26]	[27]	[28]	[29]	[30]	[31]	[32]	unpublished

[1] Non-commercial software-utility developed by Dan Schneberk of the Lawrence Livermore National Laboratory (USA) [39]. [2] Parallel research software developed by the University of Bologna (Italy) [40].

3. Results and Discussion

In this section, we present the results obtained from the tomographic observations related to the effects of time, grouping, in particular, the observed features affecting the integrity of the objects by means of different agents.

3.1. *Breaking from Prolonged Use*

A ceramic rattle ("*tintinnabulum*") from Cyprus (dated by comparative analysis back to the Iron or Roman age, 600 BCE-150 CE), modeled in the shape of a boar, has been studied with X-CT to better understand the mechanism producing the sound. The analysis actually revealed that the repeated shaking of the object had induced the formation of cracks in the ceramic material. From CT slices and a three-dimensional (3D) model, cracks are visible in the ceramic body and, even more importantly, the complete fragmentation of one of the three rattling spheres is revealed (see Figure 2). Moreover, inside one of the remaining rattling spheres a new incomplete crack is visible, suggesting that any future shaking (even only for demonstrative purposes) is to be avoided, in order to prevent further destruction.

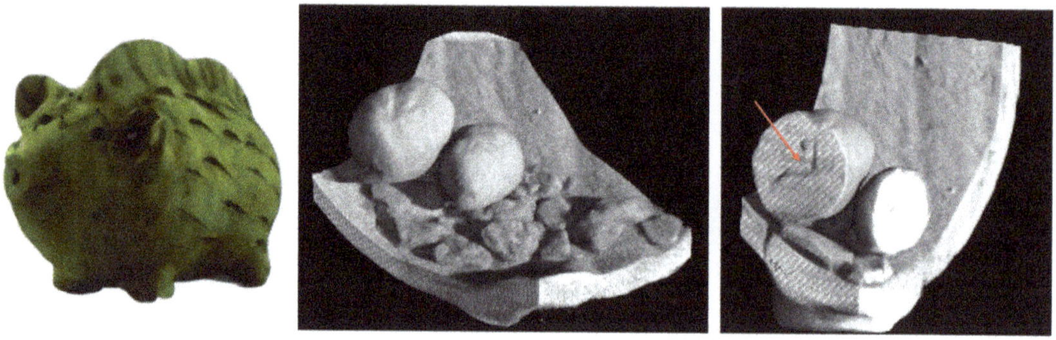

Figure 2. Examples of 3D rendering details from the X-CT analysis of the boar-shaped *tintinnabulum*, showing the fragmentation of one of the ceramic rattling spheres. The red arrow indicates an early-stage crack inside one of the remaining spheres.

3.2. *Ion Migration*

A different example is provided by a wooden-based multi-material object: a boxwood transverse flute ("Traversiere") by Lorenzo Cerino (late 18th century CE). This flute is made up of four separate pieces, and in the foot, a brass key is installed. The key was removed for the X-CT analysis, because the original scope of the investigation was to establish the conservation state (e.g., the presence of cracks) and the exact dimensions of the pieces for a metrologically accurate 3D printed replica; moreover, the metallic key might have induced artifacts in the images. Studying the foot, however, it was found that, in time, ion migration occurs inside the wood from the brass key at the installation spot: metallic residues are, in fact, visible and quite easily distinguishable in the X-CT slices as more radiopaque (brighter) areas (Figure 3).

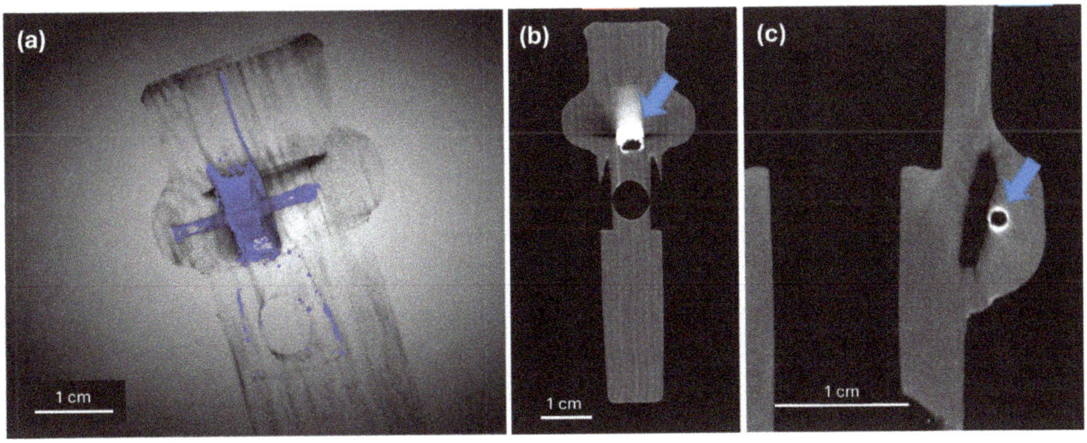

Figure 3. (**a**) 3D rendering of the Traversiere foot, where segmentation (blue) highlights the distribution of metallic residues left by the brass key. (**b**,**c**) X-CT slices with different planar orientations, where the metallic residues are identified as brighter areas around holes (blue arrows).

3.3. *Biological Activity*

The presence of xylophagous insects inside ancient wooden material is the cause of the formation of diffused holes: these can be seen sometimes from the outside on the artifact surfaces, while still not knowing the extent of the damage inside the whole wooden block.

From X-CT analysis, the extension of the insects' attack can be evaluated. During the years, it was possible for us to study some large wooden artifacts; in particular, some interesting cases are the Taiefmutmut's coffin lid (21st Dynasty, 1076–746 BCE) and the "doppio corpo" writing cabinet by Pietro Piffetti (18th century CE). In the coffin lid, despite the age of the artifact, the attack is not largely diffused but is relatively concentrated in a few areas. Performing a segmentation of the hollow parts inside the wooden material, as in Figure 4c, it is possible to visually investigate in the 3D space the distribution and shape (bending, bifurcations) of the cavities left by the insects. In Piffetti's artwork, on the other hand, the holes are located only in some of the wooden blocks used for the construction, specifically those appearing brighter in X-CT and probably made of walnut wood. The segmentation of voids in the single block revealed, in this case, a dense network of cavities (Figure 5c). Such an extensive perforation can be dangerous for the integrity of the block, also exposing the internal part of the wood to other environmental factors such as humidity and pollutants. This kind of non-invasive investigation can be a great support for conservators' interventions: repeating the analysis some time after an anoxic treatment can also be an easy way to confirm the success of the procedure.

3.4. *Environmental Effects*

The environment hosting an artifact for years, centuries or millennia can greatly influence the conditions of the object and its integrity at retrieval. Several mechanisms can be activated in time depending on the environment composition (acidic/basic), humidity conditions, mechanical stress and so on. We have been able to study several objects and different materials, each one showing different interactions and producing different stages of damage.

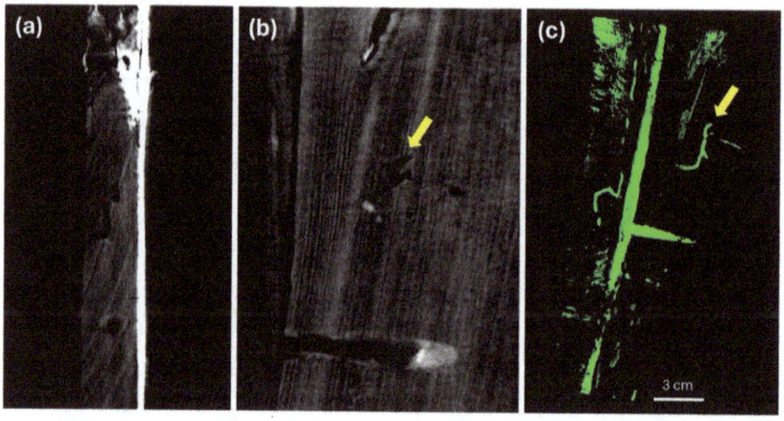

Figure 4. Lateral (**a**) and vertical (**b**) X-CT slices of the coffin lid, where different cavities produced by xylophagous insects are visible. (**c**) 3D rendering with a segmentation (green) of the hollow parts in the wood, explicating the complete structure of the cavities: as reference, the yellow arrows indicate the same feature in single slice and 3D volume.

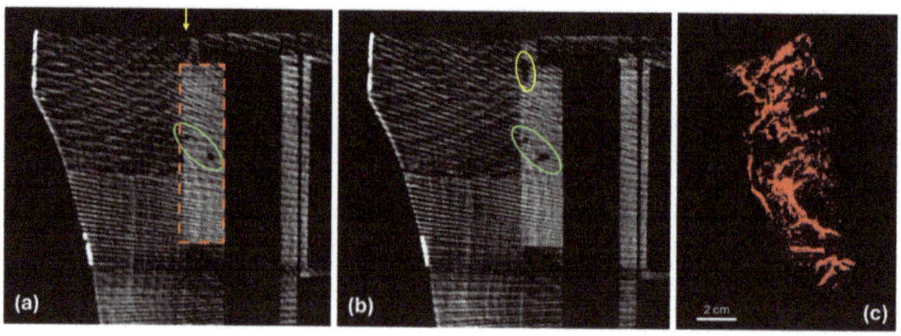

Figure 5. (**a**,**b**), X-CT horizontal slices of the "doppio corpo" cabinet (specifically the fall-front desk), where green and yellow circles/arrows highlight different cavities produced by xylophagous insects. The orange dashed rectangle in (**a**) identifies instead the wooden block represented in the 3D reconstruction (**c**) where the whole network of cavities was segmented and shown three-dimensionally in orange hue.

The first example is a soil block extracted directly from the excavation terrain at the necropolis of Villalfonsina (Abruzzo, Italy), probably dating back to the 6th–4th centuries BCE. From the preliminary observations, the bronze artifact embedded in soil was already clearly identified as fragile and fragmented. In fact, the bronze belt contained in the block was exposed during burial both to the chemical effects of organic decomposition of the body originally wearing the ornament and to the pressure of the burial soil, causing the flaking and splintering of the metal into small fragments.

CT segmentation revealed in a non-invasive way the distribution, shape and dimensions of fragments, isolating them from the earthy material (as shown in Figure 6), as well as indications about the belt's general conditions, the identification of repairs done in the past and the presence of different materials. This information supported the belt restoration in accounting for the documentation of all the fragments to be extracted and cleaned and their original position for solving the puzzle. The belt was finally reassembled and protected using resins and waxes.

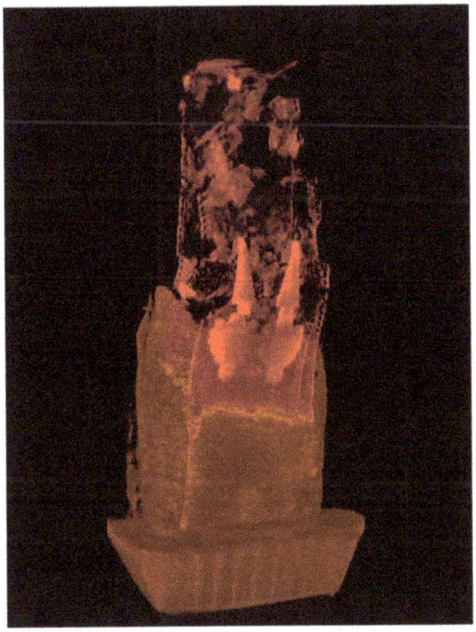

Figure 6. The X-CT segmentation helped in the virtual isolation of the bronze belt fragments from the soil block retrieved from the archaeological site. In this way it is possible to extract the fragments in their exact position (in red) from the soil (in light brown).

Another example is provided by archaeological glass kept in the ground until the present day. The glass fragments presented here were discovered on the plow soil surface around the ancient Roman city of Aquileia (North-Eastern Italy) and are part of a larger set identified to thoroughly study the different types of corrosion that can occur on the vitreous material after the long interaction with soil. These two fragments in particular are characterized by the formation of a crack network (n. 581755) and the presence of a superficial iridescent patina (n. 581681). μ-X-CT analysis allowed the determination of the size of cracks and pits and their extension inside the glass, as well as the morphology of the patina and its stratification.

The cracks in sample 581755 feature rounded ends, suggesting the possible influence of soil mechanics in fracturing the glass through erosion. From reconstruction and segmentation, it was possible to visualize the dense crack network and some larger and smaller soil grains entrapped in the main hole (the sample was originally a bead; see Figure 7). The cracks extend deep into the object (even 4–5 mm) with diameters of about 100–200 μm. Inside the cracks, it is visible from CT slices the presence of a material less radiopaque than the glass bulk (Figure 7d): it was possible to estimate that more than 10% of the glass object is altered into this different material, possibly hydrated silica.

For sample 581681 it was instead possible to measure the thickness of the patina (almost 2 mm thick) and to observe that it is composed of a sequence of alteration packets and air, apparently detaching from the bulk while following the same profile. The sample also shows the presence of pits (as the one shown in Figure 8b): the high resolution of μ-CT allows us to identify in the pit an ordered structure of concentric layers. All these phenomena are in general due to cycles of dissolution and reprecipitation of the glass material, exacerbated by the presence of salts in the soil.

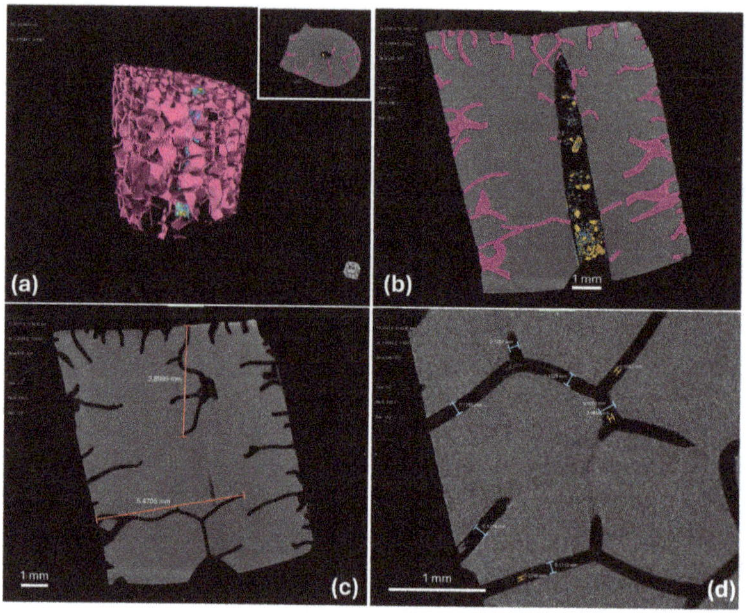

Figure 7. Glass sample from Aquileia n. 581755. (**a**) 3D segmentation of cracks (purple) and larger (yellow) and smaller (cyan) soil grains trapped in the hole; a horizontal section is reported in the top right for comparison; (**b**) vertical section with segmentations, showing the distribution of grains in the hole; a measure of (**c**) the highest penetration of cracks inside the glass and (**d**) the diameter of single cracks with (light blue) and without (orange) considering the alteration layer of hydrated silica.

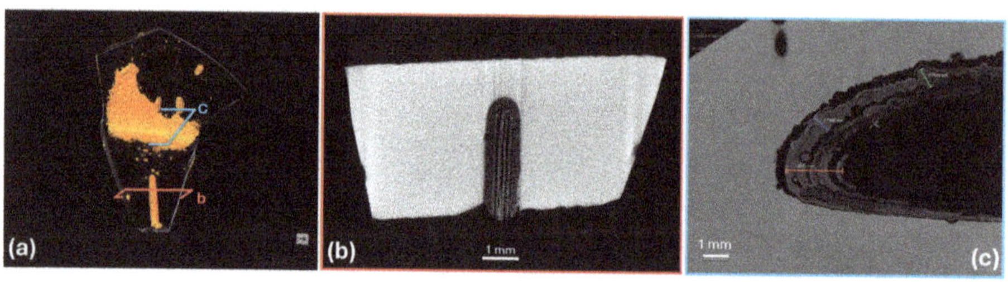

Figure 8. Glass sample from Aquileia n. 581681. (**a**) 3D segmentation of the iridescent patina (orange) in the 3D reconstruction of the fragment. (**b**) horizontal section of the vertical part of fragment (red plane in (**a**)), where the section of the widest pit is visible; (**c**) lateral section of the fragment (blue plane in (**a**)) in the area of the patina, highlighting that the patina itself is composed of multiple layers.

μ-CT resolutions are even able to discriminate features in very small artifacts, such as small carved beads. Within a study on lapis lazuli provenance, μ-CT was performed on three lapis lazuli beads from the Royal Graves of Ur (Mesopotamia, Iraq, 3rd millennium BCE). One of the issues of archaeological lapis lazuli is the frequent alteration of the pyrite crystals present inside this kind of semi-precious stone, with a higher occurrence on the surfaces. The alteration is caused by environmental conditions such as high humidity, percolating waters, or thermal stress, and it results in the loss of sulfur and in the consequent transformation of pyrite into iron oxide–hydroxide minerals. It was verified that the alterations are recognizable with μ-CT thanks to the different radiopacity, resulting from the loss of S and the remaining Fe and O as main elements. The extension of this modification is also evaluable in terms of thickness and diffusion of the alteration layers in the whole volume. It is interesting to notice that the tomographic investigation revealed the presence of some cracks in the bead (orange arrows in Figure 9), which may be the vector for humidity coming from the external environment to reach the inner pyrite crystals, inducing the alteration also in the deepest crystals.

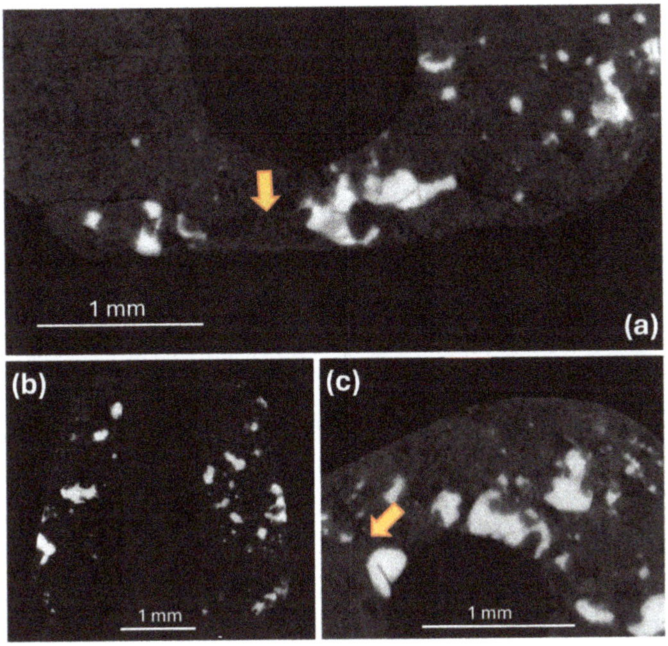

Figure 9. μ-CT of small lapis lazuli beads. (**a**,**c**) horizontal sections, (**b**) vertical section. The brighter areas are pyrite crystals, where it is also visible the few-micrometers thick alteration layer. Orange arrows indicate some cracks in the bead that may bring humidity from the external environment to the inner pyrite crystals, inducing the alteration.

Lastly, we present here the first insights that can be obtained from the μ-CT analysis of a bronze arrowhead unearthed from the burial chamber of the Urama-chausuyama mounded tomb (end of the 3rd century CE). As shown in Figure 10a,b, the arrowhead presents a black/dark green appearance with different hues, with a quite homogeneous surface apart from some localized material losses and cracks (see Figure 1). As reported in the literature, in high-tin bronze, as in the case under investigation, the dark patina could be formed by a mineralization process with a loss of copper and lead [34]. It could act as a kind of natural protective, as in the case of ancient mirrors [35], because it is stronger in resisting corrosion processes [36]. This is not the only possible mechanism of mineralization, but, in general, the lower density and atomic weight of the corroded material is an indicator of a degradation process of the bronze. This can be well highlighted by means of μ-CT (Figure 10c), observing the areas with lower absorption of X-rays (dark areas). In particular, we can see that the darker areas are present not only on the surface that was in contact with the soil, but also widely spread below the patina, particularly starting from the fractures and losses in the patina itself. In some areas, the corrosion crosses the arrowhead from side to side (red and yellow slices in Figure 10c). The actual state of preservation of the pristine metal can be evaluated by performing a segmentation of the most absorbing volumes (brighter in CT), presumably the healthiest parts of the arrowhead. The result is presented in Figure 10d and it shows a very different situation from what was deduced from optical observations, where the artifact appears to be in better health. Information obtained via μ-CT, in addition to providing insight into the state of preservation of the arrowhead, can also be used to consider appropriate conservation measures and to verify whether the process of degradation is still ongoing or has stopped, performing additional μ-CT analyses in the future.

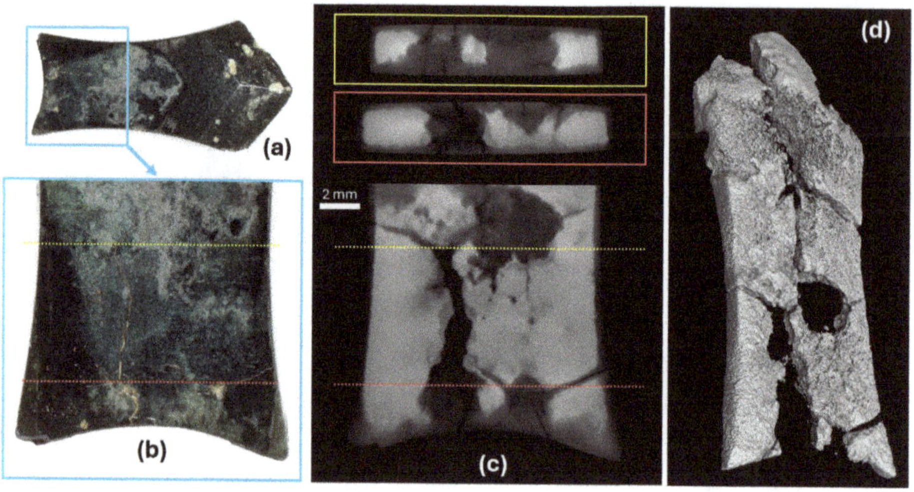

Figure 10. (**a**) The bronze arrowhead from the Urama-chausuyama mounded tomb; (**b**) magnification of the lower part of the arrowhead. (**c**) X-CT horizontal and vertical sections of (**b**); corroded areas (darker regions) are diffused on the surface and below the surface at the fractures sites, but not visible under optical inspection. (**d**) 3D rendering of the whole arrowhead where segmentation of the pristine metal helps to understand the real state of conservation of the object.

4. Conclusions

The different case studies presented in this work highlight how X-CT can reveal a variety of modifications induced in the whole volumes of artifacts by their conservation history and environment. Sometimes, surprisingly, the effects can have a high relevance on the future health of the objects, and they should be taken into consideration by the restorers who are devoted to their preservation. X-CT should be more and more considered and exploited as a support in conservation plans and strategies, in order to better target the interventions and also avoid unnecessary operations that would just put more stress on the object. As shown above, X-CT setups can be versatile and the best one in terms of energy, resolution, time and costs can be selected to better adapt to the conservation questions and the object itself. Together with microclimatic and environmental studies, X-CT can be fundamental in promoting preventive conservation and protecting our Cultural Heritage from further damage while maintaining its valorization.

Author Contributions: Conceptualization, A.R. and A.L.G.; methodology, A.R. and A.L.G.; formal analysis, L.G., A.R., F.T., L.V. and C.R.; investigation, L.G., A.R., F.T., L.V., C.R., J.R. and A.L.G.; resources, J.R.; data curation, L.G., A.R., F.T., L.V. and C.R.; writing—original draft preparation, L.G., J.R. and A.L.G.; writing—review and editing, A.R., F.T., L.V. and C.R.; visualization, L.G., A.R., F.T., L.V., C.R. and A.L.G.; supervision, A.R. and A.L.G.; project administration, J.R.; funding acquisition, J.R. and A.L.G. All authors have read and agreed to the published version of the manuscript.

Funding: This research was partially funded by the RECTOR ("International Research Center Formation Program to Accelerate Okayama University Reform") program of Okayama University, in addition to the Innovative Knowledge Hub for Humanities and Materials Science (IKH), under the project number JPMXP1323015473, through the Ministry of Education, Culture, Sports, Science and Technology (MEXT) of Japan. This research has also been partially funded within the H2020-MSCA-RISE-2018 call with the Grant Agreement no 823826, for the project BE-ARCHAEO, "BEyond ARCHAEOlogy: an advanced approach linking East to West through science, field archaeology, interactive museum experiences". Results reported in this paper reflect only the authors view and the Agency is not responsible for any use that may be made of the information it contains. The Italian PNRR—Partenariato Esteso 5 project "CHANGES: Cultural Heritage Active Innovation for Next-Gen Sustainable Society" Spoke 6: history, conservation and restoration of Cultural Heritage is also acknowledged for financial support.

Data Availability Statement: The raw data supporting the conclusions of this article will be made available by the authors upon request.

Acknowledgments: The authors wish to acknowledge the contribution of all the people involved in the mentioned case studies previously published, who contributed with different expertise and/or provided the objects, namely: F. Albertin, C. Avataneo, R. Boano, L. Bonizzoni, M. Borla, C. Bortolin, P. Buscaglia, R. Brancaccio, L. Chiaberge, J. Corsi, G. Cotto, S. De Blasi, M. Demmelbauer, G. Dughera, E. Durisi, W. Ferrarese, G. Franceschin, A. Giovagnoli, N. Grassi, S. Grassini, C. Greco, G. Herrmann, G. Iori, R. Law, E. Longo, P. Luciani, M. Magalini, G. Mangiapane, M. Martini, P. Mereu, G. Mila, M. Nervo, N. Pastrone, F. Prino, L. Ramello, M. Ravera, G. Ricchiardi, A. Romero, R. Sacchi, A. Staiano, M. Staropoli, A. Traviglia, L. Visca, L. Zamprotta, R. Zanini. The INFN CHNet network is warmly acknowledged.

Conflicts of Interest: The authors declare no conflicts of interest. The funders had no role in the design of the study; in the collection, analyses, or interpretation of data; in the writing of the manuscript; or in the decision to publish the results.

References

1. Hughes, S. CT Scanning in Archaeology. In *Computed Tomography—Special Applications*; Saba, L., Ed.; InTech: London, UK, 2011; pp. 57–70. [CrossRef]
2. Albertin, F.; Bettuzzi, M.; Brancaccio, R.; Morigi, M.P.; Casali, F. X-Ray Computed Tomography In Situ: An Opportunity for Museums and Restoration Laboratories. *Heritage* **2019**, *2*, 2028–2038. [CrossRef]
3. Oliveira, R.; de Paula, A.; Gonçalves, F.; Bueno, R.; Calgam, T.; Azeredo, S.; Araújo, O.; Machado, A.; Anjos, M.; Lopes, R.; et al. Development and characterization of a portable CT system for wooden sculptures analysis. *Radiat. Phys. Chem.* **2022**, *200*, 110409. [CrossRef]
4. Björngrim, N.; Myronycheva, O.; Fjellström, P. The use of large-scale X-ray computed tomography for the evaluation of damaged structural elements from an old timber bridge. *Wood Mater. Sci. Eng.* **2022**, *17*, 1028–1029. [CrossRef]
5. Friml, J.; Procházková, K.; Melnykc, G.; Zikmundb, T.; Kaiser, J. Investigation of Cheb relief intarsia and the study of the technological process of its production by micro computed tomography. *J. Cult. Herit.* **2014**, *15*, 609–613. [CrossRef]
6. Vigorelli, L.; Re, A.; Buscaglia, P.; Manfredda, N.; Nervo, M.; Cavaleri, T.; Del Vesco, P.; Borla, M.; Grassini, S.; Guidorzi, L.; et al. Comparison of two ancient Egyptian Middle Kingdom statuettes from the Museo Egizio of Torino through computed tomographic measurements. *J. Archaeol. Sci. Rep.* **2022**, *44*, 103518. [CrossRef]
7. Stelzner, J.; Stelzner, I.; Martinez-Garcia, J.; Gwerder, D.; Wittköpper, M.; Muskalla, W.; Cramer, A.; Heinz, G.; Egg, M.; Schuetz, P. Stabilisation of waterlogged archaeological wood: The application of structured-light 3D scanning and micro computed tomography for analysing dimensional changes. *Herit. Sci.* **2022**, *10*, 60. [CrossRef]
8. Haneca, K.; Deforce, K.; Boone, M.N.; Van Loo, D.; Dierick, M.; Van Acker, J.; Van Den Bulcke, J. X-ray sub-micron tomography as a tool for the study of archaeological wood preserved through the corrosion of metal objects. *Archaeometry* **2012**, *54*, 893–905. [CrossRef]
9. Waltenberger, L.; Bosch, M.D.; Fritzl, M.; Gahleitner, A.; Kurzmann, C.; Piniel, M.; Salisbury, R.B.; Strnad, L.; Skerjanz, H.; Verdianu, D.; et al. More than urns: A multi-method pipeline for analyzing cremation burials. *PLoS ONE* **2023**, *18*, e0289140. [CrossRef]
10. Stelzner, I.; Stelzner, J.; Fischer, B.; Hamann, E.; Zuber, M.; Schuetz, P. A multi-technique and multiscale comparative study on the efficiency of conservation methods for the stabilisation of waterlogged archaeological pine. *Sci. Rep.* **2024**, *14*, 8681. [CrossRef]
11. Allington-Jones, L.; Clark, B.; Fernandez, V. Fool's gold, fool's paradise? Utilising X-ray micro-Computed Tomography to evaluate the effect of environmental conditions on the deterioration of pyritic fossils. *J. Inst. Conserv.* **2020**, *43*, 213–224. [CrossRef]

12. Bratasz, Ł.; Akoglu, K.G.; Kékicheff, P. Fracture saturation in paintings makes them less vulnerable to environmental variations in museums. *Herit. Sci.* **2020**, *8*, 11. [CrossRef]
13. Nicolas, T.; Gaugne, R.; Tavernier, C.; Millet, E.; Bernadet, R.; Gouranton, V. Lift the veil of the block samples from the Warcq chariot burial with 3D digital technologies. In Proceedings of the Digital Heritage 2018—3rd International Congress & Expo, San Francisco, CA, USA, 26–30 October 2018; pp. 1–8, hal-01875702.
14. Stromer, D.; Christlein, V.; Huang, X.; Zippert, P.; Hausotte, T.; Maier, A. Virtual cleaning and unwrapping of non-invasively digitized soiled bamboo scrolls. *Sci. Rep.* **2019**, *9*, 2311. [CrossRef] [PubMed]
15. Kimball, J.J.L.; With, J.; Rødsrud, C.L. A new and 'riveting' method: Micro-CT scanning for the documentation, conservation, and reconstruction of the Gjellestad ship. *J. Cult. Herit.* **2024**, *66*, 76–85. [CrossRef]
16. Buscaglia, P.; Baldi, A.; Grassini, S.; Croci, S.; Croveri, P.; Cervini, L.; Cremonini, C.; Re, A.; Lo Giudice, A. Assessment of consolidation treatments: Micro-CT as a potential tool for material's penetration detection. In Proceedings of the 2022 IMEKO TC-4 International Conference on Metrology for Archaeology and Cultural Heritage, University of Calabria, Arcavacata, Italy, 19–21 October 2022.
17. Sanches, F.A.C.R.A.; Nardes, R.C.; Santos, R.S.; Gama Filho, H.S.; Machado, A.S.; Leitao, R.G.; Leitao, C.C.G.; Calgam, T.E.; Bueno, R.; Assis, J.T.; et al. Characterization an wooden Pietà sculpture from the XVIII century using XRF and microct techniques. *Radiat. Phys. Chem.* **2023**, *202*, 110556. [CrossRef]
18. Ševčík, R.; Adámková, I.; Vopálenský, M.; Viani, A. The visualization of microcracks in aged cement mortar using tomography. *AIP Conf. Proc.* **2024**, *3094*, 500007. [CrossRef]
19. Johnson, D.; Tyldesley, J.; Lowe, T.; Withers, P.J.; Grady, M.M. Analysis of a prehistoric Egyptian iron bead with implications for the use and perception of meteorite iron in ancient Egypt. *Meteorit. Planet. Sci.* **2013**, *48*, 997–1006. [CrossRef]
20. Simon, H.J.; Cibin, G.; Reinhard, C.; Liu, Y.; Schofield, E.; Freestone, I.C. Influence of microstructure on the corrosion of archaeological iron observed using 3D synchrotron micro-tomography. *Corros. Sci.* **2019**, *159*, 108132. [CrossRef]
21. Nguyen, H.-Y.; Keating, S.; Bevan, G.; Gabov, A.; Daymond, M.; Schillinger, B.; Murray, A. Seeing through Corrosion: Using Micro-focus X-ray Computed Tomography and Neutron Computed Tomography to Digitally "Clean" Ancient Bronze Coins. *Mater. Res. Soc. Symp. Proc.* **2011**, *1319*, 305. [CrossRef]
22. Bude, R.O.; Bigelow, E.M.R. Nano-CT evaluation of totally corroded coins: A demonstration study to determine if detail might still be discernible despite the lack of internal, non-corroded, metal. *Archaeometry* **2020**, *62*, 1195–1201. [CrossRef]
23. Pagano, S.; Balassone, G.; Germinario, C.; Grifa, C.; Izzo, F.; Mercurio, M.; Munzi, P.; Pappalardo, L.; Spagnoli, E.; Verde, M.; et al. Archaeometric Characterisation and Assessment of Conservation State of Coins: The Case-Study of a Selection of Antoniniani from the Hoard of Cumae (Campania Region, Southern Italy). *Heritage* **2023**, *6*, 2038–2055. [CrossRef]
24. Abate, F.; De Bernardin, M.; Stratigaki, M.; Franceschin, G.; Albertin, F.; Bettuzzi, M.; Brancaccio, R.; Bressan, A.; Morigi, M.P.; Daniele, S.; et al. X-ray computed microtomography: A non-invasive and time-efficient method for identifying and screening Roman copper-based coins. *J. Cult. Herit.* **2024**, *66*, 436–443. [CrossRef]
25. Haubner, R.; Strobl, S. Examination of archaeological bronze parts using micro-computed tomography and metallography. *Pract. Metallogr.* **2024**, *61*, 4. [CrossRef]
26. Re, A.; Albertin, F.; Avataneo, C.; Brancaccio, R.; Corsi, J.; Cotto, G.; De Blasi, S.; Dughera, G.; Durisi, E.; Ferrrarese, W.; et al. X-ray tomography of large wooden artworks: The case study of "Doppio corpo" by Pietro Piffetti. *Herit. Sci.* **2014**, *2*, 19. [CrossRef]
27. Re, A.; Lo Giudice, A.; Nervo, M.; Buscaglia, P.; Luciani, P.; Borla, M.; Greco, C. The importance of tomography studying wooden artefacts: A comparison with radiography in the case of a coffin lid from Ancient Egypt. *Int. J. Conserv. Sci.* **2016**, *7*, 935–944.
28. Re, A.; Corsi, J.; Demmelbauer, M.; Martini, M.; Mila, G.; Ricci, C. X-ray tomography of a soil block: A useful tool for the restoration of archaeological finds. *Herit. Sci.* **2015**, *3*, 4. [CrossRef]
29. Vigorelli, L.; Re, A.; Guidorzi, L.; Brancaccio, R.; Bortolin, C.; Grassi, N.; Mila, G.; Pastrone, N.; Sacchi, R.; Grassini, S.; et al. The study of ancient archaeological finds through X-ray tomography: The case of the "Tintinnabulum" from the Museum of Anthropology and Ethnography of Torino. *J. Phys. Conf. Ser.* **2022**, *2204*, 012034. [CrossRef]
30. Magalini, M.; Guidorzi, L.; Re, A.; Tansella, F.; Picollo, F.; Sturari, S.; Aprà, P.; Herrmann, G.; Law, R.; Lemasson, Q.; et al. Micro-Computed Tomography and Micro-laser ablation on altered pyrite in lapis lazuli to enhance provenance investigation: A new methodology and its application to archaeological cases. *Eur. Phys. J. Plus* **2025**, *140*, 59. [CrossRef]
31. Zanini, R.; Franceschin, G.; Vigorelli, L.; Iori, G.; Chiaberge, L.; Longo, E.; Guidorzi, L.; Re, A.; Lo Giudice, A.; Traviglia, A. Laboratory and synchrotron X-ray computed microtomography to shed light on degradation features of corroded Roman glass. *J. Am. Ceram. Soc.* **2024**, *108*, e20241. [CrossRef]
32. Tansella, F.; Vigorelli, L.; Ricchiardi, G.; Re, A.; Bonizzoni, L.; Grassini, S.; Staropoli, M.; Lo Giudice, A. X-ray Computed Tomography Analysis of Historical Woodwind Instruments of the Late Eighteenth Century. *J. Imaging* **2022**, *8*, 260. [CrossRef]
33. Kondō, Y.; Niiro, I. (Eds.) *Urama-chausuyama Kofun [Urama-chausuyama Mounded Tomb]*; Shin'yōsha: Kyoto, Japan, 1991.
34. Meeks, M. Patination phenomena on Roman and Chinese high-tin bronze mirrors and other artefacts. In *Metal Plating and Patination*; La Niece, S., Craddock, P., Eds.; Butterworth-Heinemann: Oxford, UK, 1993.

35. Chase, W.T. Chinese Bronzes: Casting, Finishing, Patination and Corrosion. In *Ancient & Historic: Metals Conservation and Scientific Research*; Scott, D.A., Podany, J., Considine, B.B., Eds.; Getty Conservation Institute and J. Paul Getty Museum: Malibu, USA, 1995.
36. Li, B.; Jiang, X.; Tu, Y.; Fu, Q.; Pan, C. "Inward Growth" corrosion and its growth mechanism in ancient Chinese bronzes. *MRS Adv.* **2020**, *5*, 1457–1466. [CrossRef]
37. Re, A.; Brancaccio, R.; Corsi, J.; Cotto, G.; Dughera, G.; Ferrarese, W.; Grassi, N.; Lo Giudice, A.; Lusso, S.; Mereu, P.; et al. L'apparato radio-tomografico/The radio-tomographic apparatus. In *Il Progetto neu_ART. Studi e Applicazioni/Neutron and X-Ray Tomography and Imaging for Cultural Heritage, Cronache 4*; Nervo, M., Ed.; Editris: Torino, Italy, 2013.
38. Re, A.; Peruzzi, N.; Del Greco, F.; Nervo, M.; Pastrone, N.; Sacchi, R.; Lo Giudice, A. A procedure for the dynamic range characterization of X-ray imaging linear and TDI detectors. *J. Instrum.* **2024**, *19*, C10005. [CrossRef]
39. Martz, H.E.; Shull, P.J.; Schneberk, D.J.; Logan, C.M. *X-Ray Imaging*; CRC Press Inc.: Boca Raton, FL, USA, 2009.
40. Brancaccio, R.; Bettuzzi, M.; Casali, F.; Morigi, M.P.; Levi, G.; Gallo, A.; Marchetti, G.; Schneberk, D. Real-time reconstruction for 3-D CT applied to large objects of cultural heritage. *IEEE Trans. Nucl. Sci.* **2011**, *58*, 1864–1871. [CrossRef]

heritage

MDPI

Article

Radiocarbon Ages of a Possible Titian Painting

Claudio Tuniz [1,*], Fabio Marzaioli [2], Filippo Terrasi [2], Isabella Passariello [2], Giuseppe Porzio [2] and Paolo Molaro [3,4]

1 Abdus Salam International Centre for Theoretical Physics (ICTP), Strada Costiera 11, 34151 Trieste, Italy
2 Dipartimento di Matematica e Fisica (DMF), Università degli Studi della Campania "L. Vanvitelli", Viale Lincoln, 5, 81100 Caserta, Italy; fabio.marziaoli@unicampania.it (F.M.); filippo.terrasi@na.infn.it (F.T.); isabellapassariello@unicampania.it (I.P.); gporzio@na.inaf.it (G.P.)
3 INAF-Osservatorio Astronomico di Trieste, (INAF-OATs), Via G.B. Tiepolo, 11, 34143 Trieste, Italy; paolo.molaro@inaf.it
4 Institute for Fundamental Physics of the Universe (IFPU), Via Beirut 2, Miramare, 34151 Trieste, Italy
* Correspondence: ctuniz@ictp.it

Abstract: A radiocarbon study of a painting that might have been made by Titian or Tintoretto or by their workshops is presented. It could be the lost Titian self-portrait or, more convincingly, the sitter might be the Venetian Andrea Cappello, elected *Procuratore de Supra* in 1537 AD. Due to the variability in the concentration of atmospheric radiocarbon during the reference period, the calibrated ^{14}C ages of the painted canvas are 1450–1530 and 1540–1635 AD (2 σ), while those of the wooden frame are 1504–1597 and 1616–1657 AD (2 σ). They are also consistent with previous analyses based on scanning macro X-ray fluorescence. These results, combined with stylistic, considerations suggest that the painting was made in the first half of the 16th century by Titian or his workshop. Circumstantial evidence that the painting was executed between 1523 and 1528 is also discussed.

Keywords: radiocarbon dating; Titian; Jacopo Tintoretto; Jacopo Sansovino; Andrea Cappello

1. Introduction

This study is aimed at providing a time window for the execution of an old painting by means of radiocarbon dating. The painting was previously analyzed with scanning macro X-ray fluorescence (MA-XRF) using the mobile LANDIS-X instrument developed at the XRAYLab of ISPC-CNR and INFN-LNS in Catania [1]. The MA-XRF analysis showed that the pigments are consistent with paintings of the 16th century and also revealed the presence of a restoration, probably made in the 19th century [2]. The interest in this painting, shown in Figure 1, stems from the fact that the artist could be Titian or someone from his entourage, or from the workshop of Tintoretto, as suggested by the Italian Ministry of Art and Culture. In fact, the X-ray analysis revealed the notable absence of an underlying drawing, which is a characteristic of Titian's style. A resemblance with Titian's physiognomy was also noted in the sitter and it was suggested that it may be the lost self-portrait made around 1537 and mentioned by Vasari in his *Lives of the most Eminent Painters, Sculptors and Architects* [2].

However, more recently, the sitter has been more convincingly identified with the *Procuratore de Supra* Andrea Cappello (?-Venice, 1565) due to the similarity to a portrait by Jacopo Tintoretto, which belongs to the Accademia Gallery of Venice (inv. n. 121; cat n. 234; Dalla Costa, in prep). This painting shows an identical drawing of the sitter's face (Figure 2). It is interesting to note that the Tintoretto painting at the Accademia Gallery was historically attributed to Titian, and that the attribution returned to Tintoretto only in the

18th century. The file of the Accademia Gallery dealing with the Tintoretto painting reports: *the influence of the Cadore artist is clearly evident. The psychological rendering of the character is in fact of the highest quality: the Procuratore seems to slip out of the shadows and observe the spectator with a hard and inquisitive gaze and the firmness of the Cappello is further underlined by the closed hand clutching a white handkerchief that emerges from under the crimson tunic. The pictorial modeling, firm and robustly innervated, also helps to avoid any pause or ornamental underlining and to emphasize the energetic face of the Procuratore.* (Authors' translation of the text available at https://www.gallerieaccademia.it/il-procuratore-andrea-cappello, accessed on 5 January 2025).

Figure 1. The painting under study, oil on canvas 90 × 70 cm, private collection.

Figure 2. Jacopo Robusti detto Tintoretto, *Il procuratore Andrea Cappello*, olio su tela, cm 115 × 86, cat 234, © Archivio fotografico G.A.VE—"su concessione del Ministero della Cultura—Gallerie dell'Accademia di Venezia". On the right, the painting bears the Cappello coat of arms and the initial 'A' of Andrea Cappello. The similarity with the sitter of the painting shown in Figure 1 was found by Thomas Dalla Costa.

As both the artist and the sitter of the painting are unknown, the purpose of this radiocarbon study is to obtain useful elements to evaluate the extant attributions. The possible attribution to Tintoretto or Titian is very attractive since they are among the greatest painters of all times who made portraits of kings and popes and of the most important rulers of the 15th century. The state-of-the-art Accelerator Mass Spectrometry (AMS) radiocarbon dating was applied on a small fraction (tens of mg) of the specimens taken from the canvas and the frames of the painting. Radiocarbon measurements are based on the proportion of radiocarbon found in the sample and are reported in terms of radiocarbon age (RC age, *in years before present* (BP)), where *present* refers to 1950, and the half-life of ^{14}C is 5568 years assuming a constant atmospheric concentration of radiocarbon in the atmosphere. In order to derive the calendar age, the change in the atmospheric concentration of radiocarbon should be accounted for by applying calibration procedures. Radiocarbon ^{14}C ages cannot provide absolutely dated chronologies, but must be converted to calendar age equivalents using a calibration curve compensating for fluctuations in atmospheric ^{14}C concentration [3]. The main information comes from measurements on tree rings complemented by other samples of known age, which are compiled into calibration curves by the International

Calibration Group. The calibration curve is not monotonic, showing oscillations in the period relevant to this study producing multiple calendar ages for the same RC age.

This study is aimed at providing a time range for the canvas and frame to confirm that we are indeed dealing with a painting made during their lifetimes. Radiocarbon dating would either confirm or rule out the possibility that the painting was made by one of these artists or by their workshops. Titian's birth year is quite uncertain, ranging from 1477 AD to somewhere around 1490 AD, while Tintoretto was born in 1518 AD. The slight difference in their birth and the period in which they began to be active may offer some possibility of discriminating between them. Concerning the identity of the sitter, in the case of Andrea Cappello, the year of birth is not known but, his death was 1 February 1565. As discussed later, there are reasons to believe that the portrait could have been made before 1537 and possibly in the year 1523. On the other hand, if it were the lost self-portrait of Titian, it should have been painted around 1537.

2. Materials and Methods

The painting (Figure 1) comes from the collection of Giulio Parri and its story has been described in detail in [2]. A total of 4 samples were taken from the original canvas used for the painting. The original canvas appears very thin and drenched in color as it can be seen in the illustrative Figure 3. The back of the painting shown in Figure 3 reveals that the canvas was relined and the original is made of two pieces sewn together, as it was frequently made in ancient times. A sample from the canvas used for the relining was also collected. Note that both samples have been taken in the upper part of the two pieces of the canvas. The painting has a magnificent carved wooden handle frame covered by golden leaf. Two samples were taken from the main frame and one from the frame used to stretch the canvas. All samples have been numbered and placed in the plastic containers with an identification label. The photographic documentation of the sampling points is available on request. The weight of the sample material and the number of spots used are detailed in Table 1.

Figure 3. On the left: The back of the painting which shows the relined canvas and the two pieces of the original canvas sewn together. Note that the rabbet of the frame fits perfectly with the painting without signs of adjustment. The photos on the right show a sample collection from the painted canvas and the two frames.

Table 1. Samples used for the dating.

Sample Name	Sample Numbers	Mass (mg)
Painted canvas	4	360
Relining canvas	1	240
Wooden frame	2	327
Stretcher frame	1	178

One specimen of a few milligrams for each of the materials listed in Table 1 was radiocarbon dated. The analysis was performed at the CIRCE (Centre for Isotopic Research on Cultural and Environmental Heritage) Laboratory, Università degli Studi della Campania *Luigi Vanvitelli*, by means of an NEC (National Electrostatic Corporation) 3MV Tandetron-based Accelerator Mass Spectrometry System [3,4]. Before AMS analysis, samples underwent physical and chemical pretreatments with the aim of contamination removal and/or to isolate a useful carbon fraction for dating purposes. After a physical cleaning step, samples were chemically pretreated by the usual chemical processing procedure. In detail, first, α cellulose was extracted from both textile and wood specimens according to methods described in [5] to obtain more accurate results. In detail, together with the usage of typical acids and bases, a strong oxidant attack (sodium chlorite) in an acid environment was applied to samples to remove molecules other than cellulose (i.e., Lignin), potentially biasing measurements. This procedure was also tested successfully for other contaminant suppression (mostly preservatives) onto archaeological samples by the authors of [6]. Then, cellulose was combusted and graphitized by means of sealed tube combustion and zinc reduction, respectively. A sample of a sufficient mass capable to guarantee 1 mgC circa was weighed into pre-cleaned 6 mm quartz tubes with CuO, evacuated ($p < 10^{-6}$ mbar), flame-sealed, and combusted at 920 °C for 6.5 h. CO_2 produced during combustion was then cryogenically purified by other gasses and graphitized by sealing it into a Pyrex reactor, holding for Zn and TiH_2, and an inner vial holding Fe powder. Graphite was produced by letting the sealed reactors react @ 560 °C for 8 h.

Graphite specimens formed over the Fe powder during graphitization were pressed in Al cathodes and installed in the sputtering ion source of the accelerator, together with IAEA C3 [7] standard samples for normalization, OXII (NBS 4990c Oxalic Acid II) for quality check purposes, and Aesar graphite treated in the same way as other samples for background subtraction. $^{14}C/^{12}C$ isotopic ratios were measured at a terminal voltage of 2.55 MV exploiting the high-resolution capability of electrostatic and magnetic analyses performed both before and after ion acceleration. For each cathode, the raw $^{14}C/^{12}C$ ratios are given by the weighted average of 12 measurements. For unknown samples, $F^{14}C$ [8] values were obtained after normalization to the standard, background, and fractionation corrections using on-line $\delta^{13}C$ measurements. Typical 1σ uncertainties in the determination of $^{14}C/^{12}C$ isotopic ratios are of 0.3–0.5%. Finally, RC ages were expressed in years BP [8] and calibrated by means of OxCal software [9] using the INTCAL 20 [10] calibration dataset.

3. Results

Radiocarbon determinations calibrated by means of OxCal.v4.4 [9] and the IntCal20 atmospheric calibration dataset [10], together with the relative coverage for the calibration ranges, are reported in Table 2 and shown in Figure 4. Given the Suess effect on the atmospheric radiocarbon trend variation, represented by the calibration curves in the reference period (15th–17th century), the analysis formally provides two possible calendric intervals for all the samples. For the relined canvas and the the stretcher frame, the possible calendric age ranges are drastically enlarged by the Suess effect. For the painted canvas, the relative coverage probability (1σ) is of 61% for the years 1459–1521 AD and of 39% for the years 1583–1623 AD for the 16th and 17th centuries. For the wooden frame, these are of 76% for the years 1524–1572 AD and 24% for the years 1630–1645 AD, respectively. The entire period from 1459 to 1657 (2σ C.L.) is permissible for the age of all materials. However, the uncertainty in the measurement results makes it challenging to further narrow down the time interval.

Table 2. Radiocarbon determinations (errors in brackets) calibrated by means of OxCal.v4.4 [9] and the IntCal20 atmospheric calibration dataset [10]. Relative coverage for the calibration ranges are also shown.

Sample Name	Radiocarbon Age (Years)	cal AD $\pm 1\sigma$	Rel. Cov. %	cal AD $\pm 2\sigma$	Rel. Cov. %
Painted canvas	368 (40)	1459–1521	61	1450–1530	50
		1583–1623	39	1540–1635	50
Wooden frame	296 (25)	1524–1572	76	1504–1597	71
		1630–1645	24	1616–1657	29
Stretcher frame	263 (38)	1524–1560	35	1497–1601	41
		1631–1667	53	1613–1680	44
		1783–1796	12	1740–1753	2
				1763–1800	12
				1940–1944	1
Relined canvas	79 (30)	1696–1724	32	1690–1729	27
		1812–1838	29	1807–1923	71
		1878–1951	39	1947–1954	2

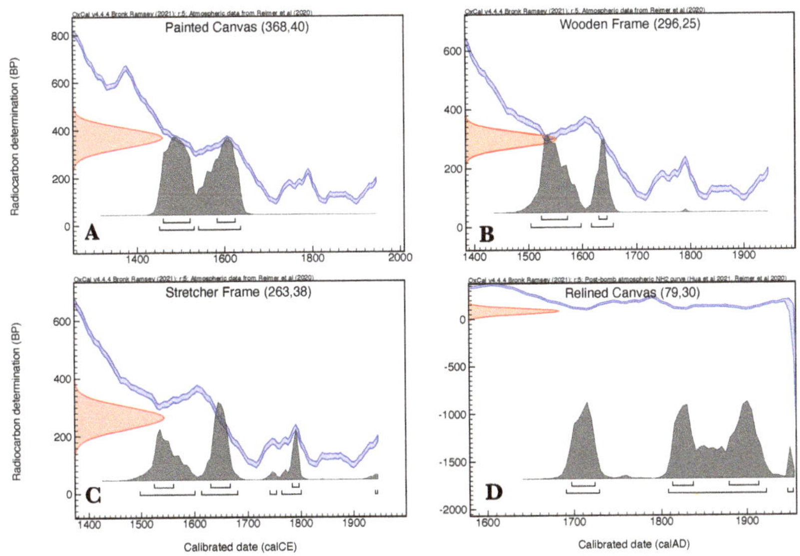

Figure 4. Observed calendric ages for the specimens from the painted canvas (**A**), the wooden frame (**B**), the stretcher frame (**C**), and the relined canvas (**D**). Measured RC ages with uncertainties are reported in brackets close to the sample name [9,10]. *X*-axis shows calendric ages curve and gray distributions show sample probability distributions.

Reported calendric ages for the main frame and the stretcher frame show that the two frames were coeval. In particular, this assures that the main frame is original and was not adapted later, for instance, at the time of relining. The relined canvas has 32% probability in 1696–1724 AD, and 29% in 1812–1838 AD with a sensitive residual probability (i.e., 39%) in the period (1878–1951 AD) due to the Suess effect. This suggests that the painting was subject to a restoration which happened somewhere later than the 18th century. Indeed, the presence of anachronistic materials were detected by MA-XRF imaging. The presence of zinc white in the collar and in an extended area of the left side of the fur and clothes, with some traces also in the beard, hair, and hands, together with the presence of cadmium red on the shoulder and on the fur, was attributed to the restoration made towards the beginning of the 20th century, probably in the period 1910–1920 [2].

4. Discussion

Overall, due to the variability in the concentration of atmospheric radiocarbon during the reference period, the radiocarbon determinations of different materials provide a chronology for the painting as being executed in a large interval between 1459 and 1657 (2σ C.L.). However, stylistic aspects and previous pigment analyses suggest that both the 15th and the 17th centuries are unlikely. In the 16th century, the probabilities for the canvas and frames for the same set of solutions overlap only in the period 1507–1530 (2σ C.L.) and, therefore, we argue that this is the more likely range for the painting's execution. These arguments, admittedly speculative, produce some elements for the attribution of the painter and on the identity of the sitter. At face value, they rule out the possibility that the author of the painting is Jacopo Tintoretto, who was born in 1518, or someone from his workshop, as suggested by the Italian Ministry of Art and Culture. On the other hand, Titian's birth, although quite uncertain, ranges from 1477 to 1490 and is more consistent with the canvas's temporal window. In the early decades of the 16th century, Titian was already a famous painter. In 1530, he painted Charles V, and three years later, he was nominated *Pintor Primero* by the Emperor and honored with the title of Count Palatine and Knight of the Golden Spur. The *Venus of Urbino* painted for Guidobaldo della Rovere is of 1538, just to mention one of the most famous ones. Considering the self-portrait hypothesis, the painting should have been made around 1537 since Vasari saw one self-portrait of Titian during his first visit of 1541 AD [2]. The period refers to the period when Titian made the paintings in the renovated church of Santo Spirito under the direction of Jacopo Sansovino, who may be connected to the frame of the painting, as we will discuss in the next section.

4.1. The Frame

The frame shown in Figure 1, which has been found to be coeval to the canvas, or slightly more recent, is particularly interesting. It is hand-sculpted by a skilled carver and gilded with real gold leaf in a style which is clearly Sansovinian. The presence of decorative elements of an architectural nature is characteristic of the new style introduced by Sansovino. Sansovino, born Jacopo Tatti (Florence, 1486–Venice, 1570), was the most important architect of the Republic of Venice since his arrival in 1527 until his death. Thanks to his friendship with Pietro Aretino and Titian, Sansovino acquired a noble status and was admitted to the Venetian patriciate, thus becoming a member of the aristocracy governing *La Serenissima*. The time range obtained for the main frame is consistent with Sansovino's arrival in Venice, and his friendship with Titian could be an indirect element in favor of the attribution of the author of the painting. It is worth noting the similarity of the frame with that of the portrait of the doctor, astronomer, and poet Girolamo Fracastoro (1476 ca.–1553) executed by Titian in 1528 on display at the National Gallery in London, and Figure 5 shows details of the two frame carriers. Both the structure of the spiral as well as the face of the little angel have common executive features that refer to a common fabrication. Since Sansovino arrived in Venice in 1527, the frame which maintains Sansovino's style would have been crafted a few years later in agreement with the radiocarbon time range for the main frame.

Figure 5. On the right, a detail of the frame of the portrait of Girolamo Fracastoro (1476 ca.–1553) executed by Titian in 1528, National Gallery, London, NG 3949, vis à vis with that of the frame under study. The similarity in the executive features suggest the execution by the same hand and likely in the same years.

4.2. The Sitter

We now consider the suggestion that the sitter is the same Andrea Cappello as in the painting by Tintoretto belonging to the Accademia Gallery of Venice, inv. n. 121; cat n. 234, shown in Figure 2 (Dalla Costa, in preparation). A direct comparison with the Tintoretto's painting of the Academia shows that both paintings, even though one sitter is depicted at bust length and the other in three quarters, are copies. The most striking difference is that while one is wearing a fur coat, the other shows the formal dress of a Procuratore da Supra. The election as *Procuratore* was a lifetime appointment in the government of the Venetian Republic and was considered second only to that of Doge. There were three categories of Procuratore, with different responsibilities and areas of the city, with the Procuratori de Supra, administering the Basilica di San Marco, being the most prestigious. The *Procuratore de Supra* wore a characteristic tunic *color cermesi*, i.e., crimson red, and is always portrayed in this way. See, for instance, the Tintoretto portraits of other *Procuratori de Supra* such as Jacopo Soranzo, Antonio Cappello or the portrait of the procurator Contarini by Domenico Tintoretto.

The two paintings differ in size. The one at the Accademia is considerably larger, measuring 115 × 86 cm. Generally, copies are smaller than the original; however, in our humble view, the one made by an anonymous painter is of comparable quality, suggesting that the artist was a very talented painter. According to what bf was reported by Francesco Sansovino, son of Jacopo, at page 238 of his book *Venetia, città nobilissima, et singolare, Descritta in XIIII libri*, published in 1581, Andrea Cappello became *Procuratore de Supra* in the year 1537 [11,12]. This is also supported by the letter by Pietro Aretino dated 22 September 1537, where he congratulates Iacopo Cornaro, Andrea Capello, and Giulio

Contarini for the election as Procuratore. A portrait of a Procuratore de Supra without the characteristic *cermesi* dress is very unlikely. For this reason, the portrait of Andrea Cappello without the red crimson dress should have been executed before the year of his appointment, namely before 1537, while the one of the Accademia would have been made after 1537. Note that the 1537 bound is marginally consistent with the temporal ranges both of the canvas and of the frames.

The Accademia Gallery reports that the painting by Tintoretto had a writing which has been removed in a recent restoration in 1959. The writing specified that the sitter was Andrea Cappello and was coming from the *Procuratoria de Supra*. Furthermore, the writing also bore the date 1523. This year remains mysterious, but we can speculate that it could be the year when Andrea Cappello was portraited alive before becoming Procuratore and copied much later by Tintoretto for the Procuratoria. It is also possible that Tintoretto made it after Cappello's death in 1565 and copied the face from the other painting. We note that a misidentification of the sitter by Tintoretto could have made it possible to mistakenly interpret the painting as a Titian self-portrait, as argued in the work of Molaro et al. [2].

5. Conclusions

Radiocarbon determinations calibrated using the code OxCal. v4.4 (Bronk Ramsey, C. 2021. OxCal v.4.4.4 URL: https://c14.arch.ox.ac.uk/oxcal.html, accessed on 5 January 2025) and the updated IntCal20 atmospheric curve [10] for different components of the canvas and the frames of a possible painting by Titian have provided two calendar solutions separated by about one century, spanning the whole period from 1459 and 1657, at 2σ C.L. However, only solutions within the 16th century are supported by stylistic arguments and by the results of the scanning macro X-ray fluorescence. Within this century, we have found the following:

- The main wooden frame provides the highest probability for the time range 1504–1597 (2σ, C.L.). The wood of the frame is handmade, beautifully carved, and covered by a golden leaf. It shows Sansovino's style and the radiocarbon date is consistent with the arrival of Jacopo Sansovino in Venice in 1527. A striking similarity of the frame with that of the portrait of Girolamo Fracastoro (1476 ca.–1553) executed by Titian in 1528 and on display at the National Gallery in London, NG 3949, is found to suggest the same hand and a similar period. The stretcher frame is also studied, providing a time range of 1497–1601 (2σ) cal AD. The stretcher frame is therefore found coeval with the main frame.
- The time range for the painted canvas falls in the period 1450–1530 cal AD, 2σ C.L. The dating of the canvas appears to be slightly prior to that of the wooden frames. However, the time range defines the manufacturing period of the canvas, which could be a few years earlier than it was painted, in particular for a canvas like this one, which was obtained by sewing two remnants.
- The relined canvas has two relatively distinct solutions for the time range, which are 1690–1729 (2σ) and 1807–1923 (2σ) with coverage probabilities of 27% and 71%, respectively. It is likely that the lining took place during the restoration of the painting estimated somewhere after the 18th century. This is consistent with the MA-XRF analysis that revealed the presence of a restoration, probably made in the 19th century [2].

Thus, the obtained radiocarbon determinations combined with stylistic considerations suggest a chronology for the painting as having being executed in the 16th century and likely before 1530. As such, the result seems to disfavor the possibility that the author of the painting is Jacopo Tintoretto or his workshop, while it is compatible with Titian or his workshop. As for the sitter, the result is consistent both with a self-portrait made by Titian around 1537 or with a portrait of Andrea Cappello made before 1537, i.e., the year when

Cappello was appointed *Procuratore de Supra*. One intriguing possibility is that the sitter is Andrea Cappello and the date of the painting is 1523, which is the year written on the portrait of Andrea Cappello made by Tintoretto and belonging to the Accademia Gallery. This is also very close to 1528, the execution year of the portrait of Fracastoro by Titian, which shows a very similar frame.

Author Contributions: C.T. contributed in the data collection, data analysis, and writing; F.M., F.T., I.P. and G.P. carried out the measurements in the laboratory; P.M. contributed to the writing of the discussion. All authors have read and agreed to the published version of the manuscript.

Funding: This research received no external funding.

Data Availability Statement: The painting studied in this article can be made available by the authors on request under a collaborative approach and permission of the owners.

Acknowledgments: We warmly thank Irka Hajdas for discussions and comments on the manuscript. We acknowledge Luba Freedman for pointing out the Sansovinian style of the frame and Thomas Dalla Costa for the convincing identification of the sitter with the Procuratore de Supra Andrea Cappello. P.M. sincerely thanks Raymond Waddington and Charles Hope for enlightening considerations about the painting. Gabriella Schiulaz is acknowledged for checking the writing.

Conflicts of Interest: The authors declare no conflicts of interest.

Abbreviations

The following abbreviations are used in this manuscript:

MA-XRF	Macro X-Ray Fluorescence
CIRCE	Center for Isotopic Research on Cultural and Environmental Heritage
NEC	National Electrostatic Corporation
BP	Before Present
RC	Radiocarbon
AMS	Accelerator Mass Spectrometry

References

1. Romano, F.P.; Caliri, C.; Nicotra, P.; Di Martino, S.; Pappalardo, L.; Rizzo, F.; Santos, H.C. Real-time elemental imaging of large dimension paintings with a novel mobile macro X-ray fluorescence (MA-XRF) scanning technique. *J. Anal. At. Spectrom.* **2017**, *32*, 773–781. [CrossRef]
2. Molaro, P.; Caliri, C.; Pavone, D.; Romano, F.P. On the Titian's self-portrait mentioned by Vasari with technical analysis of an anonymous painting. *J. Cult. Herit.* **2021**, *47*, 265–269. [CrossRef]
3. Terrasi, F.; De Cesare, N.; D'Onofrio, A.; Lubritto, C.; Marzaioli, F.; Passariello, I.; Rogalla, D.; Sabbarese, C.; Borriello, G.; Casa, G.; et al. High precision ^{14}C AMS at CIRCE. *Nucl. Instrum. Methods Phys. Res. B* **2008**, *266*, 2221–2224. [CrossRef]
4. Terrasi, F.; Rogalla, D.; De Cesare, N.; D'Onofrio, A.; Lubritto, C.; Marzaioli, F.; Passariello, I.; Rubino, M.; Sabbarese, C.; Casa, G.; et al. A new AMS facility in Caserta/Italy. *Nucl. Instrum. Methods Phys. Res. B* **2007**, *259*, 14–17. [CrossRef]
5. Passariello, I.; Marzaioli, F.; Lubritto, C.; Rubino, M.; D'Onofrio, A.; De Cesare, N.; Borriello, G.; Casa, G.; Palmieri, A.; Rugolla, D.; et al. Radiocarbon Sample Preparation at the Circe AMS Laboratory in Caserta, Italy. *Radiocarbon* **2007**, *49*, 225–232. [CrossRef]
6. Capano, M.; Marzaioli, F.; Passariello, I.; Pignatelli, O.; Martinelli, N.; Gigli, S.; Gennarelli, I.; De Cesare, N.; Terrasi, F. Preliminary Radiocarbon Analyses of Contemporaneous and Archaeological Wood from the Ansanto Valley (Southern Italy). *Radiocarbon* **2012**, *54*, 701–714. [CrossRef]
7. Rozanski, K.; Stichler, W.; Gonfiantini, R.; Scott, E.; Beukens, R.P.; Kromer, B.; Van Der Plicht, J. The IAEA 14C Intercomparison Exercise 1990. *Radiocarbon* **1992**, *34*, 506–519. [CrossRef]
8. Stuiver, M.; Polach, H.A. Discussion Reporting of 14C Data. *Radiocarbon* **1977**, *19*, 355–363. [CrossRef]
9. Bronk Ramsey, C. Oxcal v4.4., Computer Calibration Software. 2021. Available online: https://c14.arch.ox.ac.uk/oxcal.html (accessed on 5 January 2025).
10. Reimer, P.; Austin, W.; Bard, E.; Bayliss, A.; Blackwell, P.G.; Ramsey, C.B.; Butzin, M.; Cheng, H.; Edwards, R.L.; Friedrich, M.; et al. The IntCal20 Northern Hemisphere Radiocarbon Age Calibration Curve (0–55 cal kBP). *Radiocarbon* **2020**, *62*, 725–757. [CrossRef]

11. Sansovino, F. *Venetia, Città Nobilissima, et Singolare, Descritta in XIIII Libri*; Appresso Iacomo Sansovino in Venetia: Venice, Italy, 1581.
12. Rossi, P. Ritratto del Procuratore Andrea Cappello. In *Catalogo dei Ritratti Autografi*; Gallerie dell'Accademia: Venice, Italy, 1994.

MDPI AG
Grosspeteranlage 5
4052 Basel
Switzerland
Tel.: +41 61 683 77 34

Heritage Editorial Office
E-mail: heritage@mdpi.com
www.mdpi.com/journal/heritage

www.ingramcontent.com/pod-product-compliance
Lightning Source LLC
Chambersburg PA
CBHW041954260926
51850CB00019B/2710
* 9 7 8 3 7 2 5 8 7 0 9 2 9 *